T5-CAG-964

Springer Tracts in Modern Physics
Volume 158

Springer
Berlin
Heidelberg
New York
Barcelona
Hong Kong
London
Milan
Paris
Singapore
Tokyo

Springer Tracts in Modern Physics

Springer Tracts in Modern Physics provides comprehensive and critical reviews of topics of current interest in physics. The following fields are emphasized: elementary particle physics, solid-state physics, complex systems, and fundamental astrophysics.
Suitable reviews of other fields can also be accepted. The editors encourage prospective authors to correspond with them in advance of submitting an article. For reviews of topics belonging to the above mentioned fields, they should address the responsible editor, otherwise the managing editor.
See also http://www.springer.de/phys/books/stmp.html

Vladimir M. Shalaev

Nonlinear Optics of Random Media

Fractal Composites and Metal-Dielectric Films

With 51 Figures

Professor Vladimir M. Shalaev
New Mexico State University
Department of Physics, MSC 3D
P.O. Box 30001
Las Cruces, NM 88003, USA
Email: vshalaev@nmsu.edu

Physics and Astronomy Classification Scheme (PACS): 42.65.-k, 42.70.-a, 73.20.Mf, 78.30.Ly, 78.66.-w, 78.66.Sq, 81.05.Kf, 81.05.Bx, 81.05.Rm, 81.05.Ys

ISSN 0081-3869
ISBN 3-540-65615-4 Springer-Verlag Berlin Heidelberg New York

Library of Congress Cataloging-in-Publication Data.

Shalaev, Vladimir M., 1957- . Nonlinear optics of random media: fractal composites and metal-dielectric films/Vladimir M. Shalaev. p.cm. – (Springer tracts in modern physics, v. 158). Includes bibliographical references and index. ISBN 3-540-65615-4 (hardcover: alk. paper). 1. Nonlinear optics. 2. Inhomogeneous materials–Optical properties. I. Title. II. Series: Springer tracts in modern physics; 158. QC1.S797 vol. 158 [QC173.4.I53] 535'.2–dc21 99-38272

Typesetting: Data conversion by Springer-Verlag, Heidelberg
Cover design: *design& production* GmbH, Heidelberg

SPIN: 10658677 56/3144/tr - 5 4 3 2 1 0 – Printed on acid-free paper

To my son Ilia

Preface

Professor Ping Sheng, who became one of the protagonists in the physics of random inhomogeneous media, told me once that, when he started research in this field thirty years ago, his senior colleagues tried to discourage him from studying "dirty" materials and suggested he focused on traditional "nice" solids, such as crystals. Nowadays, it has become clear that disordered materials may possess unique physical properties that are significantly enhanced in comparison with their geometrically ordered counterparts.

Among the traditional problems of the physics of random media are electron and photon localization, wave and particle transport through disordered systems, percolation, coherence and nonlinear effects. While there are well established theoretical approaches for the case of linear transport and optical properties, the nonlinear phenomena are still waiting for a complete theoretical description.

In this book, I review recent advances in the nonlinear optics of random media. Significant progress in this field became possible, to a large extent, because of the breakthrough in understanding and modeling irregular structures of complex disordered systems. Among basic models for the description of various random composites are fractal structures, self-affine surfaces, and percolation systems. These three models well simulate a large class of random media, including nanostructured composite materials, rough thin films, clusters, cermets, colloidal aggregates, porous media, and many others.

We consider here random media that consist of very small structural units (particles or roughness features). By "small" particles we mean those whose size is much less than the wavelength λ of the applied field, so that the quasistatic approximation can be used to describe the response of an individual particle. Typically, the particle size ranges from ten to a hundred nanometers. The size of a whole system, in general, can be arbitrary with respect to the wavelength λ of the light wave.

Nonlinear optical phenomena experience strong enhancement at resonance with the system's "internal" fundamental frequencies. At resonance, it is possible to obtain strong nonlinearities that are of special interest for various applications. In this book, we focus on the *resonance* nonlinear optics of random media.

We assume that each particle, or roughness feature, constituting a random medium may exhibit the resonance optical response. Mainly, we consider metallic particles, where the resonance is associated with collective electron oscillations called surface plasmons. The displacement of free electrons from their equilibrium position in a small particle results in a noncompensated charge on the surface of the particle, leading to its polarization; this polarization, in turn, results in a restoring force that causes electron oscillations. The light-induced oscillating dipoles of different particles interact with each other, forming collective optical excitations of the whole system. The particles are embedded in a host medium and they can form objects of complex geometry, such as fractals, self-affine films, and percolation metal-dielectric films. Such objects are considered in this book.

We should note that although most of the theoretical models and experiments considered here deal with complex systems of metal nanoparticles, the developed formalism is general and it can be applied to other particles possessing optical resonance. For instance, it can be used for a system of semiconductor particles where the resonance is typically associated with excitons. Also, most of the theoretical approaches described, such as the one involving the coupled-dipole equations (Chaps. 3 and 4), can be applied to an arbitrary system of particles, random or not; thus the models described here can be used for a broad class of media.

The problem under consideration hereafter is how the collective optical response of a complex random system of particles can be expressed in terms of the optical properties of individual particles that are assumed to be known. In other words, we want to learn how the global morphology (i.e. geometry) of a nanostructured random medium affects its collective optical excitations.

It is well known in physics that dynamical excitations of a system strongly depend on the system's symmetry. For example, all physical excitations of a translationally invariant crystal, such as phonons, polaritons, magnons, etc., are represented by running plane waves, or, mathematically, by the Fourier harmonics. This is because the Fourier harmonics are the eigenfunctions of the shift operator, ∇, that represents the translational invariance symmetry.

The random media that we consider here have a different symmetry. They are often scale-invariant within a certain interval of sizes, i.e. they look self-similar in different scales. The key approach to describing such symmetry is the concept of fractals. With some variations, this symmetry manifests itself in all the randomly structured materials considered here: small-particle aggregates, self-affine surfaces and percolation metal-dielectric films.

Science has always striven to reduce complex problems to their simplest components; fractal geometry is one such appealing approach to describing a broad class of random media. A different symmetry associated with the fractal geometry results in unique physical properties significantly differing from "conventional" crystals, with their translational invariance symmetry. One of the most important effects of the fractal morphology is localization of dynamical excitations, including the optical excitations considered in this book. The localization of optical excitations in nanometer-sized areas of a random medium results in "hot spots", areas where the electromagnetic energy is accumulated or one can say focused within the sub-wavelength parts of the systems, breaking in some sense the diffraction limit of the conventional far-zone optics. The large local fields in the hot spots result eventually in giant enhancement of optical nonlinearities, a phenomenon that is also considered in the book.

In summary, the optics of random media display a rich variety of effects, and some of these effects are hardly intuitive. Field localization of various sorts occur and recur in a wide gamut of disordered systems, most strikingly in those possessing dilational symmetry, leading to the enhancement of many optical phenomena, especially nonlinear processes. Making judicious use of these enhancement effects, as well as of other aspects of the many complex resonances that distinguish these systems, can lead to new and unexpected physics and many applications. When developed, in the fullness of time, these disordered materials may attain a level of practical importance and versatility that might rival or surpass their geometrically ordered counterparts.

Finally, I need to say a few words to guide you on the contents of the book. Chapter 1 introduces basic models of irregular structures of random inhomogeneous media and briefly reviews the main approaches used to describe their *linear* electromagnetic properties. Chapter 2 discusses the origin of the enhanced optical nonlinearities in random media and introduces basic formulas for calculation of the enhancement. Nonlinear optics of various random media, namely small-particle fractal aggregates, self-affine surfaces and metal-dielectric films, are considered in Chaps. 3, 4, and 5, respectively. Chapters can be read to a large extent independently, so that a reader who is interested in a specific problem can start by looking directly at the relevant chapter.

It is important to note that theories and experiments described in the central chapters of the book (Chaps. 3, 4, and 5) originate from collaborative studies with a number of excellent researchers I was privileged to work with during the last ten years. I would like especially to acknowledge: Prof. M. I. Stockman and Drs. V. A. Markel, R. Botet, and E. Y. Poliakov, each of whom contributed significantly to developing the optical theory of fractals and self-affine surfaces; Prof. A. K. Sarychev, who played a key role in building a theory of nonlinear optical properties of percolation systems; Mr. V. A. Shubin and Mr. V. A. Podolskiy, who lately helped to develop further the optics of metal-dielectric films. In my work, I always feel the inspiring influence of experimentalists and, especially, my friends Prof. M. Moskovits and Dr. V. P. Safonov, who are making a difference to the experimental optics of fractals, as well as Dr. S. Grésillon and Profs. P. Gadenne, A. C. Boccara and J. C. Rivoal, who together with their colleagues performed important experimental studies of metal-dielectric films. I would also like to thank Profs. A.

K. Popov, G. zu Putlitz, R. Jullien, T. F. George, R. K. Chang, P. Sheng, A. A. Maradudin, R. W. Boyd, R. L. Armstrong, S. G. Rautian, V. V. Slabko, S. I. Bozhevolnyi, D. J. Bergman, D. Stroud, and Drs. E. B. Stechel and V. Y. Yakhnin for continual support of my work and many fruitful discussions.

Las Cruces, New Mexico, USA, June 1999 *Vladimir M. Shalaev*

Contents

1. Electromagnetic Properties of Random Composites. A Review of Basic Approaches

"Would you tell me, please, which way I ought to go from here?"
"That depends a good deal on where you want to get to," said the Cat.
"I don't much care where –" said Alice.
"Then it doesn't matter which way you go," said the Cat.
"– so long as I get somewhere," Alice added as an explanation.
"Oh, you're sure to do that," said the Cat, "if you only walk long enough."

Lewis Carroll, *Alice's Adventures in Wonderland*

1.1 Introduction to Random Media: Fractals, Self-Affine Surfaces and Percolation Composites

Electromagnetic phenomena in random metal-insulator composites, such as rough thin films, cermets, colloidal aggregates and others, have been intensively studied for the last two decades [1]. These media typically include small nanometer-scale particles or roughness features. Nanostructured composites possess fascinating electromagnetic properties, which differ greatly from those of ordinary bulk material, and they are likely to become ever more important with the miniaturization of electronic and optoelectronic components.

Recently, new efficient models for description of irregular and inhomogeneous structures of random media have been developed; among them are fractal structures, self-affine surfaces and various percolation models.

Often random nanocomposites and rough thin films, within certain intervals of size, are characterized by a fractal, i.e. scale-invariant, structure. Fractals look similar in different scales; in other words, a part of the object resembles the whole. Regardless of the size, this resemblance persists for ever, if the object is mathematically defined. In Nature, however, the scale-invariance range is restricted, on the one side by the size of the structural units (e.g. atoms) and on the other by the size of the object itself. The emergence of fractal geometry was a significant breakthrough in the description of irregularity [2, 3]. The realization of the fact that the geometry of fractional dimensions is often more successful than Euclidean geometry in describing natural shapes and phenomena provided a major impetus in research and led to a better understanding of many processes in physics and other sciences.

Fractal objects do not possess translational invariance and therefore cannot transmit running waves [3, 4]. Accordingly, dynamical excitations, such as, for example, vibrational modes (fractons), tend to be localized in fractals [3–6]. Formally, this is a consequence of the fact that plane running waves are not eigenfunctions of the operator of dilation symmetry characterizing fractals. The efficiency of fractal structures in damping running waves is probably the key to a "self-stabilization" of many of the fractals found in nature [3].

The physical and geometrical properties of fractal clusters has attracted researchers' growing attention in the past two decades [2–7]. The reason for this is twofold. First, processes of aggregation of small particles in many cases lead to the formation of fractal clusters rather than regular structures [2, 3, 7]. Examples include the aggregation of colloidal particles in solutions [8], the formation of fractal soot from little carbon spherules in the process of incomplete combustion of carbohydrates [9], the growth of self-affine films [10] (see below), and the gelation process and formations of porous media [2, 3]. The second reason is that the properties of fractal clusters are very rich in physics and different from those of either bulk material or isolated particles (monomers) [11].

The most simple and extensively used model for fractal clusters is a collection of identical spherically symmetrical particles (monomers) that form a self-supporting geometrical structure. It is convenient to think of the monomers as of identical rigid spheres that form a bond on contact. A cluster is considered self-supporting if each monomer is attached to the rest of the cluster by one or more bonds. Fractal clusters are classified as "geometrical" (built as a result of a deterministic iteration process) or "random". Most clusters in nature are random.

The major models for computer simulation of random fractal clusters are the cluster-cluster aggregation model [7] and the diffusion-limited aggregation (Witten-Sander) model [12]. The majority of natural clusters can be accurately described by one of these models, with some variations [7].

The fractal (Hausdorff) dimension of a cluster D is determined through the relation between the number of particles N in a cluster (the aggregate) and the cluster's gyration radius R_{c}:

$$N = (R_{\mathrm{c}}/R_0)^D \ , \tag{1.1}$$

where R_0 is a constant of the order of the minimum separation distance between monomers. Note that the fractal dimension is, in general, fractional and less than the dimension of the embedding space d, i.e. $D < d$. Such a power-law dependence of N on R_{c} implies a spatial scale-invariance (self-similarity) for the system. For the sake of brevity, we refer to fractal aggregates, or clusters, as fractals.

Another definition of the fractal dimension utilizes the pair density-density correlation function $\langle \rho(\mathbf{r})\rho(\mathbf{r}+\mathbf{R})\rangle$:

$$\langle \rho(\mathbf{r})\rho(\mathbf{r}+\mathbf{R})\rangle \propto R^{D-d}, \quad \text{if } R_0 \ll r \ll R_c \ . \tag{1.2}$$

This correlation makes fractals different from truly random systems, such as salt scattered on the top of a desk. Note that the correlation becomes constant, $g(r) = \text{const}$, when $D = d$; this corresponds to conventional media, such as crystals, gases and liquids. The unusual morphology associated with fractional dimensions results in unique physical properties of fractals, including the localization of dynamical excitations.

As shown in the following chapters, optical excitations in fractal composites are substantially different from those in other media. For example, there is only one dipolar eigenstate that can be excited by a homogeneous field in a dielectric sphere (for a spheroid, there are three resonances with non-zero total dipole moment); the total dipole moment of all other eigenstates is zero and, therefore, they can be excited only by an inhomogeneous field. In contrast, fractal aggregates possess a variety of dipolar eigenmodes, distributed over a wide spectral range, which can be excited by a homogeneous field.

In the case of continuous media, dipolar eigenstates (polaritons) are running plane waves that are eigenfuctions of the operator of translational symmetry. This also holds in most cases for microscopically disordered media that are homogeneous on average. Dipolar excitations in these cases are typically delocalized over large areas, and all monomers absorb light energy at approximately the same rate in regions that significantly exceed the wavelength. In contrast, fractal composites have optical excitations that are localized in small sub-wavelength regions. The local fields in these "hot" spots are large and the absorption by the "hot" monomers is much higher than by other monomers in a fractal composite. This is a consequence of the fact (mentioned above) that fractals do not possess translational symmetry; instead, they are symmetrical with respect to the scale transformation. We consider the optical properties of small-particle fractals in Chapter 3 of the book.

Fractal nanostructured materials can be fabricated with the aid of well established chemical and depositional methods. For example, colloidal clusters with the fractal dimension $D = 1.78$ can be grown in colloidal solutions via the cluster-cluster aggregation process [7, 8], and clusters with fractal dimension $D = 2.5$ can be grown by the particle-cluster aggregation process (Witten-Sander aggregation, WSA, [7, 12]).

An important special case of fractals is a self-affine structure. In contrast to conventional fractals, self-affine films have scaling properties in the (x, y) plane different from those in the direction z normal to the plane [10] (see Chap. 4). Self-affine thin films with various fractal dimensions may be grown by controlling conditions of atomic beam deposition and substrate temperature [10]. Note that a deposition of colloidal aggregates originally prepared in solution also results in self-affine structures. We consider the optical properties of self-affine thin films in Chap. 4.

Another important model used for the description of random composites, such as metal-dielectric films (also referred to as semicontinuous metal films), is the percolation model, which is closely related to the concept of fractals

[13, 14]. These films can be produced, for example, by metal evaporation or sputtering onto an insulating substrate.

Percolation represents probably the simplest example of a disordered system. Consider a square lattice, where each site is occupied randomly with probability p (or is empty with probability $1 - p$). Assume that occupied sites imply electrical conductors, empty sites represent insulators, and electrical current can flow only between nearest-neighbor conductor sites. Then, there is a critical (threshold) concentration p_c above which the dc current can flow (percolate) from one edge of the lattice to the other; this is the so-called "site percolation." When the bonds between the sites are randomly occupied, we speak of "bond percolation." The most common example of bond percolation is a random resistor network, where the metallic wires in a regular network are cut randomly with probability $q = 1 - p$. Again, there is a critical density $q_c = 1 - p_c$ that separates a conductive phase at low q from an insulating phase at large q. Perhaps the most natural example of percolation is continuum percolation, such as a sheet of conductive material with circular holes punched randomly in it (the Swiss cheese model). In contrast to site or bond percolation, in continuum percolation the positions of the two components of a random mixture (in this case, the presence or absence of holes) are not restricted to the discrete sites of a regular lattice.

In percolation, the concentration p to some extent plays a similar role as the temperature in thermal phase transitions: long-range correlations control the percolation transition and the relevant quantities near p_c are described by power laws with some critical exponents.

A percolation system can be thought of as a set of clusters (consisting of connected bonds). For $p < p_c$, only finite clusters exist; at $p = p_c$ there appears an infinite cluster. The mean size of the finite clusters, for p below and above p_c, is characterized by the correlation length ξ ($\sim R_c$) that increases as $\xi \sim |p - p_c|^{-\nu}$ when p approaches p_c. As was first pointed out by Stanley [14], these finite clusters can be described as fractals for $r \ll \xi$. Thus, the number of sites (bonds) in the percolation clusters is $N \sim r^D$ for $r \ll \xi$ and $N \sim r^d$ for $r \gg \xi$.

Near the percolation (conduction) threshold, the scaling theory can be applied to describe the electromagnetic response. We consider briefly the basic concepts of ac conductivity theory for percolation systems in Sect. 1.4 of this chapter. In Chap. 5, using the percolation model, we develop a theory of the nonlinear optical response in random metal-dielectric films.

The electromagnetic response of random media can be generally described in terms of the complex dielectric function (permittivity) $\epsilon \equiv \epsilon' + i\epsilon''$, or complex conductivity $\sigma \equiv \sigma' + i\sigma''$; these two quantities are related through the known formula $\epsilon = 4\pi i\sigma/\omega$. If particles in a cluster are conductive and connected, there is a flow of electrons ("Ohmic current") through the system. Alternatively, there is a dipolar response, which arises in Maxwell's equations through the displacement field (polarization). Formally, one can say that the

Ohmic current dominates in the low-frequency region (when $|\epsilon'| \ll \epsilon''$), and the displacement field (dipolar response) dominates in the high-frequency region (when $|\epsilon'| \gg \epsilon''$). We note, however, that introducing the complex permittivity (or, equally, complex conductivity) unifies the concepts of currents and dipoles by making them simply two different approaches for the same electromagnetic response.

In the following sections of this chapter we will consider some basic models and approaches, such as mean-field approaches (Sect. 1.2), spectral representation (Sect. 1.3), and the percolation theory (Sect. 1.4). These models are traditionally used to describe *linear* electromagnetic and optical properties of random nanostructured media. Models recently developed for *nonlinear* optical phenomena in various random media will be considered in the following chapters of the book.

1.2 Mean-Field Theories and Numerical Techniques

One of the appealing features of mean-field and effective-medium theories is the ease with which one may calculate the dielectric constant of a composite material $\epsilon_{\rm e}$. We note, however, that despite the fact that mean-field theories are often successful in describing the linear optical properties of random composites, their use for the nonlinear optical effects is limited. This is because the mean field theories do not properly treat the local field fluctuations that play a crucial role in the nonlinear optical response of random media, as will be shown in the following chapters of the book.

In the case of a two-phase d-dimensional medium, the Maxwell-Garnett theory (MGT) yields the following expression [15] for the effective dielectric constant $\epsilon_{\rm e}$ in terms of the dielectric constants of host medium ϵ_2 and spherical inclusions ϵ_1 (present with volume fraction p_1):

$$\frac{\epsilon_{\rm e}-\epsilon_2}{\epsilon_{\rm e}+(d-1)\epsilon_2} = p_1\frac{\epsilon_1-\epsilon_2}{\epsilon_1+(d-1)\epsilon_2}. \tag{1.3}$$

(Note that similar approaches have been also developed earlier for dielectrics by Clausius [16] and Mossotti [17], and applied in optics by H. Lorentz [18] and L. Lorentz [19].) The MGT expression is obviously nonsymmetrical with respect to the exchange $\epsilon_1 \to \epsilon_2, \epsilon_2 \to \epsilon_1$ and is justified only in the limit of small p_1 when it can be simplified as:

$$\epsilon_{\rm e} = \epsilon_2 + 3p_1\epsilon_2\frac{\epsilon_1-\epsilon_2}{\epsilon_1+(d-1)\epsilon_2} + O(p^2). \tag{1.4}$$

Thus, in the dilute limit, where $p \ll 1$, the interaction between particles is small and there is only one resonance at $\epsilon_1 = -2\epsilon_2$ (for $d = 3$), corresponding to the surface plasmon resonance of an isolated spherical particle. For metal particles in a vacuum, in accordance with (1.3), the resonance occurs at $\omega = \omega_p/\sqrt{3}$.

Note that the plasmon resonance depends on the restoring force resulting from a noncompensated surface charge; this force in turn depends on the shape of a particle. Therefore, we can say that the plasmon resonance is a geometrical resonance, with the resonance frequency depending on the particle's shape.

The absorption coefficient $\alpha = 2(\omega/c)\mathrm{Im}\sqrt{\epsilon_e}$ for $p \ll 1$ is given by (after setting $\epsilon_1 = \epsilon$ and $\epsilon_2 = 1$)

$$\alpha \approx 3p_1 \frac{\omega}{c} Im \left[\frac{\epsilon - 1}{\epsilon + 2} \right]. \tag{1.5}$$

Figure 1.1 shows the absorption coefficient of a dilute suspension of metal spheres in vacuum, as calculated from (1.5) [20]. The surface plasmon resonance results in a strong absorption near $\omega = \omega_p/\sqrt{3}$. In the limit $\omega\tau \ll 1$, MGT gives an ω^2 dependence for the absorption:

$$\alpha = C\omega^2 p_1, \text{ where } C = 9/[4\pi c\sigma(0)], \tag{1.6}$$

where $\sigma(0)$ is the static conductivity. While the experiment does show the predicted dependences on ω^2 and p_1, the magnitude of the absorption by a composite is typically much larger than that predicted by MGT. Enhancement of the far-infrared absorption by a composite can be explained by interactions between the particles and will be discussed in Chap. 3.

For larger values of p_1, (1.3) becomes a poor approximation. In particular, it fails to have a nontrivial percolation threshold for either of the two phases. The symmetric (in the two components) effective-medium theory

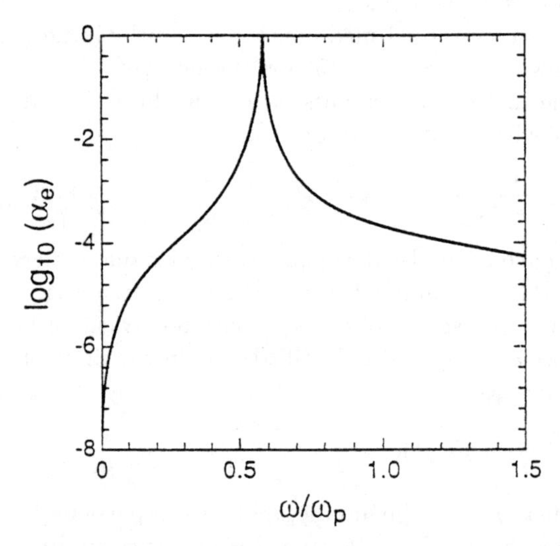

Fig. 1.1. Absorption coefficient $\alpha \equiv \alpha_e$ for a composite of volume fraction 0.01 of spheres of a Drude metal ($\omega_p\tau = 100$) embedded in a host medium of dielectric constant unity, as calculated in the quasi-static approximation and the dilute limit [20]

(EMT), known also as coherent potential approximation, was first proposed by Bruggeman [21], and it offers the following formula for calculating ϵ_e

$$p_1 \frac{\epsilon_1 - \epsilon_e}{\epsilon_1 + (d-1)\epsilon_e} + p_2 \frac{\epsilon_2 - \epsilon_e}{\epsilon_2 + (d-1)\epsilon_e} = 0. \tag{1.7}$$

This is a quadratic equation with the solution (see, for example, [22])

$$\epsilon_e = \frac{1}{2(d-1)} \{ d\bar{\epsilon} - \epsilon_1 - \epsilon_2 \pm [(d\bar{\epsilon} - \epsilon_1 - \epsilon_2)^2 + 4(d-1)\epsilon_1\epsilon_2]^{1/2} \}, \tag{1.8}$$

where $\bar{\epsilon} \equiv p_1\epsilon_1 + p_2\epsilon_2$. The upper sign in (1.8) should be used when ϵ_1 and ϵ_2 are both real and positive.

Figure 1.2 shows $\mathrm{Re}[\sigma_e(\omega)] = (\omega/4\pi)\mathrm{Im}[\epsilon_e(\omega)]$, plotted against frequency for several values of $p = p_1$, as calculated within the EMT [23]. In contrast to a sharp peak described by the MGT expression, the EMT shows, for $p < p_c$ ($p_c = 1/3$ in the EMT), a single peak that is broadened by electromagnetic interactions between individual grains. For $p > p_c$, a Drude peak, centered at $\omega = 0$ and corresponding to the dc conductivity of the composite, develops in addition to the surface plasmon band. The integrated strength of

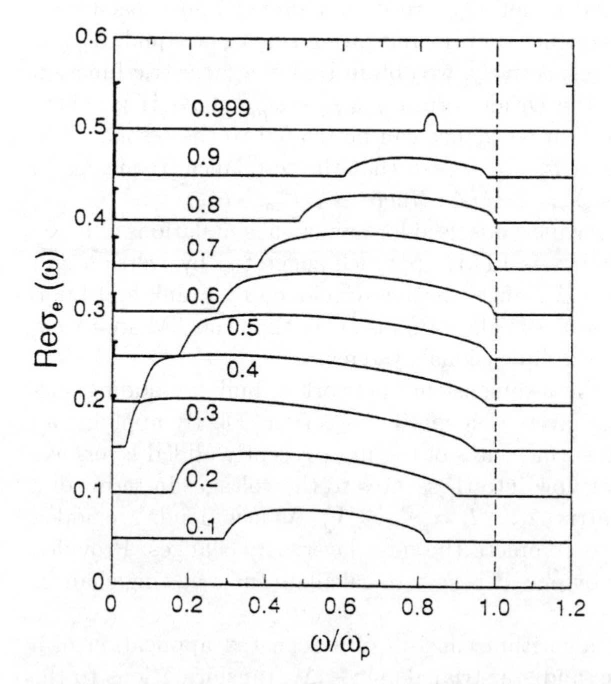

Fig. 1.2. Schematic of $\mathrm{Re}[\sigma_e(\omega)]$ of a metal-insulator composite made up of volume fraction p of a Drude metal and $1-p$ of insulator, as calculated in the EMT in the limit $\tau \to \infty$. The heavy vertical line at $\omega = 0$ denotes a δ-function, which represents the Drude peak; the integrated strength of the δ-function is proportional to the height of the δ-function. The peak at $p = 0.999$ is arbitrarily increased in height for clarity [23]

the Drude peak grows as p increases. The surface plasmon band eventually shrinks and narrows to a peak, centered at $\omega = \omega_p\sqrt{2/3}$, that corresponds to a void resonance (a charge oscillation in the vicinity of a spherical void in an otherwise homogeneous metal). Frequency dependences similar to those shown in Fig. 1.2 have been seen in experiments [24].

We briefly mention below the numerical techniques used for calculation of electromagnetic response of random composites.

The random resistor (R) and resistor-inductor-capacitor (RLC) network models are widely used to describe electromagnetic properties of a percolation system; they permit the study of dc and ac conductivity respectively [25, 26]. A network containing complex conductivities of two kinds, chosen to represent "insulating" and "conducting" particles, can describe composites – for instance, of a Drude metal and dielectric.

An insulating bond in this model is represented by a capacitor $C_{\rm i}$ with conductivity $\sigma_{\rm i} = -{\rm i}\omega C_{\rm i}$, so that the corresponding dielectric constant for an insulator (dielectric) is $\epsilon_{\rm i} = 4\pi C_{\rm i}$. A metallic bond is represented as a series of a resistor and an inductor in parallel with a capacitor. Then the conductivity of a metallic bond is $\sigma_{\rm m} = (R - {\rm i}\omega L)^{-1} - {\rm i}\omega C_{\rm m}$, where R is the resistance of the conducting element and L and C_m are its inductance and capacitance. By setting the plasmon frequency and the relaxation rate to be equal to $\omega_p = (4\pi/L)^{1/2}$ and $\Gamma = R/L$, respectively, we obtain that the dielectric function for the metallic bond has the Drude form $\epsilon_{\rm m} = \epsilon_0 - \omega_p^2/[\omega(\omega + {\rm i}\Gamma)]$, where $\epsilon_0 = 4\pi C_{\rm m}$. Then the plasmon resonance can be related to the resonance in the LC- circuit occurring at $\sigma_{\rm m} = -\sigma_{\rm i}$, so that the resonance frequency, in the limit of small losses is $\omega_r = 1/\sqrt{LC}$, where $C = C_{\rm m} + C_{\rm i}$.

Among the most effective methods used for numerical simulations of RLC-networks are the transfer-matrix (TM) approach suggested by Derrida and coworkers [27] and the $Y - \Delta$ transformation developed by Frank and Lobb [28]. Note that although the $Y - \Delta$ algorithm is faster than the TM approach, it can be applied only to two-dimensional systems.

In the TM approach, the d-dimensional network is built by adding successive $(d-1)$-dimensional layers in a specific direction [20]. By applying an arbitrary voltage V_j at all surface sites of the most recently added layer, one relates the currents I_i that flow into these sites to the voltages in terms of a symmetric admittance matrix $A_{ij} : I_i = \sum_j A_{ij} V_j$. As new bonds are added to the network in order to complete the next layer, A_{ij} changes. Provided new bonds are added one by one, it is easy to calculate the resulting changes in A_{ij}.

The Frank and Lobb algorithm consists of a repeated application of a sequence of series, parallel and star-triangle $(Y - \Delta)$ transformations to the bonds of the lattice. The final result of this sequence of transformations is to reduce any finite portion of the lattice to a single bond that has the same conductance as the entire lattice.

A review of the calculation results for the optical properties of random composites, using the above numerical algorithms, can be found, e.g. in [11].

1.3 Spectral Representation

Spectral representation is a very efficient approach for calculation of the dielectric function in inhomogeneous media. It was developed by Bergman [22, 29], Fuchs et al. [30–34] and by Milton [35]. Below we briefly consider the basic idea of the spectral theory following a paper by Sheng and others [36] that provides a simple description of the theory.

We consider small particles (e.g. metal particles) with dielectric function ϵ embedded in a host medium characterized by dielectric constant ϵ_i. In the quasi-static approximation, the problem to be solved is represented by the equation

$$\boldsymbol{\nabla} \cdot \left[1 - \frac{1}{s}\eta(\mathbf{r})\right] \boldsymbol{\nabla}\phi = 0, \tag{1.9}$$

where $s = \epsilon_\mathrm{i}/(\epsilon_\mathrm{i} - \epsilon)$ is the only parameter that characterizes the material properties of the system. Here ϕ is the electrical potential and $\eta(\mathbf{r})$ is defined as having the value 1 at those points where the polarizable particles are located, and zero otherwise. The formal solution to the above equation at the condition $\Delta\phi/l = E = 1$ for the applied field polarized along, say, the z axis, can be expressed in operator notations as

$$\phi = -\frac{z}{1 - \hat{\Gamma}/s}, \tag{1.10}$$

where

$$\hat{\Gamma} = \frac{1}{V} \int \mathrm{d}\mathbf{r}' \eta(\mathbf{r}') \boldsymbol{\nabla}' G_0(\mathbf{r} - \mathbf{r}') \cdot \boldsymbol{\nabla}' \tag{1.11}$$

is an integral–differential operator, with $G_0(\mathbf{r} - \mathbf{r}') = 1/4\pi|\mathbf{r} - \mathbf{r}'|$ being the Green function for the Laplace equation, and V being the sample volume. By defining the inner product operation as

$$\langle\phi|\psi\rangle = \int \mathrm{d}\mathbf{r}' \eta(\mathbf{r}') \boldsymbol{\nabla}'\phi^* \cdot \boldsymbol{\nabla}'\psi, \tag{1.12}$$

we can write the effective dielectric function (to be exact, its diagonal (z, z) component) as

$$\frac{\epsilon_\mathrm{e}}{\epsilon_\mathrm{i}} = -\frac{1}{V} \int \mathrm{d}\mathbf{r} \left(1 - \frac{\eta}{s}\right) \frac{\partial\phi}{\partial z} = 1 + \frac{1}{s}\langle z|\phi\rangle \frac{1}{V}. \tag{1.13}$$

From (1.10) and (1.13), we find that the spectral representation (also referred to as the Bergman–Milton representation) for the effective dielectric function is

$$\frac{\epsilon_\mathrm{e}}{\epsilon_\mathrm{i}} = 1 - \frac{1}{V} \sum_n \frac{|\langle z|\phi_n\rangle|^2}{s - s_n} = 1 - \sum_n \frac{f_n}{s - s_n} \tag{1.14}$$

or, in the explicit form,

$$\frac{\epsilon_{\rm e}}{\epsilon_{\rm i}} = 1 - \frac{1}{V} \sum_{n,m} \langle z|\psi_n\rangle \langle \psi_n|(s-\hat{\Gamma})^{-1}|\psi_m\rangle \langle \psi_m|z\rangle, \tag{1.15}$$

where s_n and ϕ_n are the nth eigenvalue and eigenfunction of the operator $\hat{\Gamma}$, and $\{\psi_n\}$ is an arbitrary complete basis set.

Note that in the spectral representation (1.14) the information about the structure is separated from the information about the material. The microstructural information is given by the spectral function (for the z-component), i.e. by $f_n = |\langle z|\phi_n\rangle|^2$ and the location of the poles s_n. The poles s_n must lie in the interval between 0 and 1 and can be interpreted as depolarization factors for different eigenmodes that are characterized by their strength f_n [22, 29, 33, 34].

We should also note that the spectral representation expresses the effective dielectric function of a composite in terms of the mode strengths f_n (and, for a continuous medium, in terms of $g(s) = \sum_n f_n \delta(s - s_n)$) but it does not provide a method for determining the mode strength and density from first principles. If however, the effective dielectric function is known – for example, it is given by the Maxwell-Garnett theory or by the effective-medium theory considered above – then the mode density can be found explicitly [31].

In Chap. 3, we consider a microscopic approach based on solving the coupled-dipole equations (CDE) and expressing the optical properties in terms of the eigenfunctions and eigenvalues of the operator of the dipole-dipole interaction between the particles. This method is similar in many respects to the spectral representation considered above but, in contrast to the latter, it provides the means from first principles for calculating mode strength and density.

1.4 Critical Behavior of Conductivity and Dielectric Function in the Percolation Model

In this section we briefly review the basic results of the percolation theory for *linear* electromagnetic responses of a random composite. In Chap. 5 we generalize the scaling approach of the percolation theory for *nonlinear* optical responses.

Pioneering work on electromagnetic properties of a percolation system has been carried out by Efros and Shklovskii [37] and by Straley [38]. Using the scaling approach from a phase-transition theory, they have developed a theory of electrical transport in a metal-insulator composite near a percolation threshold.

In the quasi-static limit, the problems of finding the electrical conductivity and the dielectric function are equivalent, since the corresponding equations for the current density and conductivity σ, and for the displacement field and dielectric function ϵ, are identical. Accordingly, electrical and dielectric properties of inhomogeneous media can be equally described in terms of either the complex conductivity, $\sigma \equiv \sigma' + \mathrm{i}\sigma''$, or the complex dielectric function, $\epsilon \equiv \epsilon' + \mathrm{i}\epsilon''$; as mentioned above, these two quantities are related via the equation $\epsilon = (4\pi\mathrm{i}/\omega)\sigma$.

For a Drude metal, the dielectric constant is given by

$$\epsilon = \frac{4\pi\mathrm{i}\sigma}{\omega} = \epsilon_0 + \frac{4\pi\mathrm{i}\sigma(0)}{\omega[1 - i\omega\tau]}, \tag{1.16}$$

where dc conductivity $\sigma(0)$ is related to plasma frequency ω_p and relaxation time τ by $\sigma(0) = \omega_p^2\tau/(4\pi)$, and ϵ_0 is the contribution to ϵ due to interband electron transitions. Note that for the relaxation rate of collective plasmon oscillation $1/\tau$ the following different notations $1/\tau = \omega_\tau = \Gamma$ are interchangeably used in the literature and throughout this book.

The Drude model describes well the optical response associated with free electrons in metals; through the term ϵ_0 it also takes into account the contribution to the dielectric constant due to interband electron transitions. The real and imaginary parts of the Drude dielectric function can be also represented as

$$\epsilon' = \epsilon_0 - \frac{\lambda^2}{\lambda_p^2}\frac{1}{1 + (\lambda/\lambda_\tau)^2}, \tag{1.17}$$

and

$$\epsilon'' = \frac{\lambda^3}{\lambda_p^2\lambda_\tau}\frac{1}{1 + (\lambda/\lambda_\tau)^2}, \tag{1.18}$$

where $\lambda/\lambda_\tau \equiv (\omega\tau)^{-1}$ and $\lambda/\lambda_p \equiv (\omega_p/\omega)$.

In the low-frequency limit, when $\epsilon'' \gg \epsilon'$ (i.e. $\sigma' \gg \sigma''$), the inequality $\omega\tau \ll 1$ is valid for a Drude metal and the dielectric function is approximated by

$$\epsilon' = \epsilon_0 - 4\pi\sigma(0)\tau = \epsilon_0 - \lambda_\tau^2/\lambda_p^2, \qquad \epsilon'' = 4\pi\sigma(0)/\omega = \lambda_\tau\lambda/\lambda_p^2. \tag{1.19}$$

We assume that $|p - p_\mathrm{c}| \ll p_\mathrm{c}$ and $H \equiv \epsilon_\mathrm{i}/\epsilon \ll 1$, where ϵ_i and ϵ are dielectric functions for the (host) insulator and metal constituents respectively. If these two requirements are met, one can apply the theory of [37, 38]. The parameter H plays in this theory a role similar to a magnetic field in the ferromagnetic phase transition theory [37].

The effective dielectric function ϵ_e of a composite material near a percolation threshold can be described by the following general formula [37–40]

$$\frac{\epsilon_\mathrm{e}}{\epsilon} \sim L^{-\frac{t}{\nu}} F\left(\frac{\epsilon_\mathrm{i}}{\epsilon} L^{(t+s)/\nu}\right), \tag{1.20}$$

where length scale L is given by the minimum of all the scales:

$$L = \min\{l, \xi, L_\omega\}\,. \tag{1.21}$$

Here l is the linear size of a system, ξ is the percolation correlation length, and L_ω is the coherence length, which can often be identified with the localization length. (All lengths are measured in units of a typical grain size a, which is much less than the light wavelength.)

The scaling function $F(z)$ in (1.20) has the limiting behavior described next [37–40].

For large values of $|z|$, $F(z)$ is given by

$$F(z) = A_0 z^{t/(t+s)}, \qquad |z| \gg 1. \tag{1.22}$$

Thus, it does not depend on length scale L.

For small $|z|$, the result depends upon whether there exists a conducting path connecting opposite sides of the sample. If such a path exists, i.e. if $p > p_c$,

$$F(z) = A_1 + A_2 z, \qquad |z| \ll 1, \quad (p > p_c). \tag{1.23}$$

If there is no a conducting path across the sample and the sample is insulating for a dc signal,

$$F(z) = A_3 z + A_4 z^2, \qquad |z| \ll 1, \quad (p < p_c). \tag{1.24}$$

Since $\epsilon_i \ll \epsilon$, the condition $|z| \ll 1$ corresponds to relatively small scales L, for which there is a distinction between conducting and insulating parts of a system, whereas at very large scales, $|z| \gg 1$, the existence of a conducting part becomes unimportant, and therefore $F(z)$ has no dependence on L and its form is the same for $p > p_c$ and $p < p_c$.

For $|z| \gg 1$, we obtain from (1.20) and (1.22) the following result for ϵ_e which is independent of the length scale:

$$\epsilon_e/\epsilon \sim (\epsilon_i/\epsilon)^{t/(t+s)}, \qquad |z| \gg 1, \tag{1.25}$$

for both $p > p_c$ and $p < p_c$. This result is usually associated with the anomalous frequency dependence (see also below) and corresponds to the case of a high magnetic field in the ferromagnetic phase-transition theory.

If $l \ll L_\omega$, ξ, i.e. the length scale of importance is the size of the system l, then from the above relations we obtain

$$\epsilon_e \sim l^{-t/\nu} \epsilon \tag{1.26}$$

for a conducting system and

$$\epsilon_e \sim l^{s/\nu} \epsilon_i \tag{1.27}$$

for a non-conducting system.

Below we assume that l exceeds both ξ and L_ω, i.e. one of the two latter lengths defines the system behavior, according to (1.20) and (1.21). We analyze the dielectric function ϵ_e for the two limiting cases, $\xi \ll L_\omega$ and $\xi \gg L_\omega$. First, we consider the case when percolation correlation length ξ is small, i.e. $\xi \ll L_\omega$.

The percolation correlation length ξ has the following critical behavior [13]

$$\xi \sim |p - p_c|^{-\nu}. \tag{1.28}$$

As mentioned above, the function $F(z)$ defined in (1.22) results in $\epsilon_{\rm e}$ given by (1.25), which is independent of L (and therefore ξ), in the limit $|z| \gg 1$.

In the other limit $|z| \ll 1$ and $|p - p_{\rm c}| \to 1$, we see from (1.20), (1.23) and (1.24) that $\epsilon_{\rm e} \to \epsilon$ and $\epsilon_{\rm e} \to \epsilon_{\rm i}$ for $p > p_{\rm c}$ and $p < p_{\rm c}$, respectively. This agrees, of course, with the expected limiting behavior.

In accordance with (1.20), (1.22)–(1.24) and (1.28), the real part of the dielectric function has the following behavior in the considered case when ξ is the smallest length scale in (1.21) [39]:

$$\epsilon_{\rm e}' = \begin{cases} A_0 \epsilon_{\rm i}^{t/(t+s)} [\omega/4\pi\sigma(0)]^{-s/(t+s)} \cos(\frac{\pi}{2}\frac{s}{t+s}), & \text{if } |z| \gg 1 \\ A_2 \epsilon_{\rm i} |p - p_{\rm c}|^{-s}, & \text{if } |z| \ll 1,\, p > p_{\rm c} \\ A_3 \epsilon_{\rm i} |p - p_{\rm c}|^{-s}, & \text{if } |z| \ll 1,\, p < p_{\rm c}. \end{cases} \tag{1.29}$$

According to (1.29), in the limit $|z| \ll 1$ the quantity $\epsilon_{\rm e}'$ has the same scaling dependence (as a function of $p - p_{\rm c}$) both below and above the threshold [37]. Note also that $\epsilon_{\rm e}'$ has a peak at $\omega = 0$, as a function of frequency. The half-width of the peak, $\Delta\omega \approx [4\pi\sigma(0)/\epsilon_{\rm i}]|p - p_{\rm c}|^{t+s}$, decreases to zero at $p_{\rm c}$, and its height is proportional to $\epsilon_{\rm i}|p - p_{\rm c}|^{-s}$ (i.e. it diverges at $p_{\rm c}$ [39]).

For the imaginary part of the dielectric function, $\epsilon_{\rm e}'' = (4\pi/\omega)\sigma_{\rm e}'$, in the considered case $L = \xi$, we obtain from (1.20), (1.22)–(1.24) and (1.28) [37, 38, 41, 42]

$$\epsilon_{\rm e}'' = \begin{cases} A_0 \epsilon_{\rm i}^{t/(t+s)} [\omega/4\pi\sigma(0)]^{-s/(t+s)} \sin(\frac{\pi}{2}\frac{s}{t+s}), & \text{if } |z| \gg 1 \\ A_1 4\pi\sigma(0)\omega^{-1}|p - p_{\rm c}|^{t}, & \text{if } |z| \ll 1,\, p > p_{\rm c} \\ -A_4 \epsilon_{\rm i}^2 [\frac{\omega}{4\pi\sigma(0)}] |p - p_{\rm c}|^{-t-2s}, & \text{if } |z| \ll 1,\, p < p_{\rm c}, \end{cases} \tag{1.30}$$

where $A_4 < 0$ in (1.30).

The absorption coefficient is defined by $\alpha = 2(\omega/c){\rm Im}\sqrt{\epsilon_{\rm e}}$. In the low-frequency limit ($\omega\tau \ll 1$), we find $\epsilon \approx {\rm i}\epsilon'' \approx {\rm i}\omega_p^2\tau/\omega$. Then, for $|z| \ll 1$ and $p < p_{\rm c}$, using (1.29) and (1.30) we find $\epsilon_{\rm e}'' \ll |\epsilon_{\rm e}'|$, $\epsilon_{\rm e}' \sim z$, and $\epsilon_{\rm e}'' \sim z^2$. Accordingly, $\alpha \approx (\omega/c)\epsilon_{\rm e}''/\sqrt{\epsilon_{\rm e}'}$, and [41]

$$\alpha \sim \frac{\epsilon_{\rm i}^{3/2}}{\omega_p^2 \tau c}\omega^2 |p - p_{\rm c}|^{-(t+3s/2)}. \tag{1.31}$$

Equation (1.31) gives the known quadratic frequency-dependence for α. (For a Drude metal, this dependence is also predicted by the effective-medium theory in the dilute limit, as shown above in Sect. 1.2.)

For $|z| \ll 1$ and $p > p_{\rm c}$, comparing (1.29) and (1.30) we obtain that $\epsilon_{\rm e}'' \gg |\epsilon_{\rm e}'|$ since $\epsilon_{\rm e}' \sim O(1)$ and $\epsilon_{\rm e}'' \sim z$. Thus $\alpha \approx \sqrt{2}(\omega/c)\sqrt{\epsilon_{\rm e}''}$ and [41]

$$\alpha \sim (\omega_p/c)(\omega\tau)^{1/2}|p - p_{\rm c}|^{t/2}. \tag{1.32}$$

Equation (1.32) represents the $\omega^{1/2}$ dependence of the Hagen-Rubens relation for conducting materials.

For $|z| \gg 1$, the absorption coefficient shows the anomalous frequency dependence [see (1.25)]

$$\alpha \sim c^{-1}(\omega_p^2\tau)^{s/2(s+t)}\epsilon_{\mathrm{i}}^{t/2(s+t)}\omega^{(s+2t)/2(s+t)}. \tag{1.33}$$

In this limit, the ac conductivity has the following frequency dependence [37, 39, 43]:

$$\sigma_{\mathrm{e}}' = (\omega/4\pi)\epsilon_{\mathrm{e}}'' \propto \omega^{t/(s+t)}.$$

We now consider the other limiting case, when $L_\omega \ll \xi$, so that $L = L_\omega$ in (1.21). The condition $L_\omega \ll \xi$ means that during a period $\sim \omega^{-1}$ a random walker traverses a region smaller than the correlation length ξ (and all other length scales considered above).

The excitation (coherence) length L_ω can be related to the mean-square distance traveled in a random walk with travel time t by the expression [4–6, 44]

$$[\langle r^2(t)\rangle]^{1/2} \propto t^{1/(2+\Theta)}, \tag{1.34}$$

where $2+\Theta = d_{\mathrm{w}} = 2D/\tilde{d}$ is the fractal dimension of the random walk. The exponent $\tilde{d}$ is the fracton (spectral) dimension that determines the spectral dependence of the density of vibrational states (fractons) as follows [5, 6]

$$\rho \sim \omega^{\tilde{d}-1}. \tag{1.35}$$

For homogeneous media, $\Theta = 0$ and (1.34) gives the usual diffusion law. For fractal clusters, $\Theta > 0$ reflects the slowing down of the diffusion process in fractals [5, 6, 44].

The frequency ω of the applied field may be used to determine the travel time t during which the random walk traverses the region L_ω according to

$$L_\omega \propto \omega^{-1/(2+\Theta)}. \tag{1.36}$$

For the case of $|z| \ll 1$, we obtain from (1.20) and (1.23) the result $\epsilon_{\mathrm{e}} \sim \epsilon L_\omega^{-t/\nu}$, which with the use of (1.36) leads to $\epsilon_{\mathrm{e}}'' \propto \omega^{[t/\nu(2+\Theta)]-1}$. The corresponding ac conductivity is given by

$$\sigma_{\mathrm{e}}' \propto \omega^{[t/\nu(2+\Theta)]}.$$

This result was first reported by Gefen, Aharony and Alexander [44], who developed a theory of anomalous diffusion on percolation clusters. Their method consisted of integrating (1.34) with the cluster-size distribution in a percolation system. Note that, as was pointed out by the authors [44], this method does not take into account capacitances between different clusters and so assumes that there is no interaction between the clusters of a percolation system.

To summarize, (1.20)–(1.24) allow us to calculate the dielectric function (and conductivity) of a percolation system in all limiting cases, $|z| \gg 1$ and $|z| \ll 1$ (corresponding to high and low "magnetic fields"), with $p > p_c$ and $p < p_c$ (metallic and dielectric behavior, respectively).

In Chap. 5, we develop the percolation theory to include nonlinear optical responses of random composites.

2. Surface-Enhanced Nonlinear Optical Phenomena

All the fifty years of conscious brooding have brought me no closer to the answer to the question, "What are light quanta?" Of course today every rascal thinks he knows the answer, but he is deluding himself.

Albert Einstein

2.1 Introduction

In this chapter, physical reasons for enhancement of nonlinear optical processes in random media are briefly outlined and basic calculation formulas are introduced. In later chapters of the book, the enhanced optical phenomena will be considered in more detail for different nanostructured materials.

Giant enhancement of optical responses in a random medium including a metal component, such as metal nanocomposites and metal rough thin films consisting of small nanometer-sized particles or roughness features, is associated with optical excitation of surface plasmons that are collective electromagnetic modes and strongly depend on the geometrical structure of the medium. As discussed in Chap. 1, nanocomposites and rough thin films are, typically, characterized by fractal geometry, where collective optical excitations, such as surface plasmons, tend to be localized in small nm-sized areas, namely hot spots; this is because the plane running waves are not eigenfunctions of the operator of dilation symmetry that characterizes fractals.

Thus, in fractals collective plasmon oscillations are strongly affected by the fractal morphology, leading to the existence of hot and cold spots (i.e. areas of high and low local fields). Local enhancements in the hot spots can exceed the average surface enhancement by many orders of magnitude because the local peaks of the enhancement are spatially separated by distances much larger than the peak sizes. The spatial distribution of these high-field regions is very sensitive to the frequency and polarization of the applied field [11, 45–50]. The positions of the hot spots change chaotically but reproducibly with frequency and/or polarization. This is similar to speckles created by laser light scattered from a rough surface, with the important difference that the scale-size for fractal plasmons in the hot spots is in the nanometer range rather than in the micrometer range encountered for photons.

Two classes of surface plasmons are commonly recognized: localized surface plasmons (LSPs) for individual particles, and surface plasmon waves (SPWs) occurring on a relatively smooth metal surface. SPWs are also called surface plasmon polaritons (SPPs), which are coherent mixtures of plasmons and photons. SPWs propagate laterally along a metal surface, whereas LSPs are confined to metal particles. However, in fractal media, plasmon oscillations in different particles strongly interact with each other via dipolar or, more generally, multipolar forces. Thus, plasmon oscillations in random objects, such as fractal aggregates, self-affine surfaces and metal-dielectric films, are neither conventional SPWs nor independent LSPs. Rather, they should be treated theoretically as collective optical modes resulting from interactions between the optically excited individual particles of the object.

The fractal plasmon, as any wave, is scattered from density fluctuations – in other words, fluctuations of polarization. The strongest scattering occurs from inhomogeneities of the same scale as the wavelength. In this case, interference in the process of multiple scattering results in Anderson localization. The Anderson localization corresponds typically to uncorrelated disorder. A fractal structure is in some sense disordered, but it is also correlated for all length scales, from the size of constituent particles, in the lower limit, to the total size of the fractal, in the upper limit. Thus, what is unique for fractals is that, because of their scale invariance, there is no characteristic size of inhomogeneity – inhomogeneities of all sizes are present, from the lower to the upper limit. Therefore, whatever the plasmon wavelength, there are always fluctuations in a fractal with similar sizes, so that the plasmon is always strongly scattered and, consequently, can be localized [11, 46].

Because of a random character of fractal surfaces, the high local fields associated with the hot spots look like strong spatial fluctuations. Since a nonlinear optical process is proportional to the local fields raised to a power greater than one, the resulting enhancement associated with the fluctuation area (i.e. with the hot spot) can be extremely large. In a sense, we can say that enhancement of optical nonlinearities is especially large in fractals because of very strong field fluctuations.

Large fluctuations of local electromagnetic fields on a metal surface of inhomogeneous metal media result in a number of enhanced optical effects. A well known effect is surface-enhanced Raman scattering (SERS) by molecules adsorbed on a rough metal surface, e.g. in aggregated colloid particles [51, 52]. Giant enhancements of *nonlinear* optical responses in metal fractals have also been predicted in early paper [45].

In an intense electromagnetic field, a dipole moment induced in a particle can be expanded into a power series: $d = \alpha^{(1)}E(r)+\alpha^{(2)}[E(r)]^2+\alpha^{(3)}[E(r)]^3+ ...$, where $\alpha^{(1)}$ is the linear polarizability of a particle, $\alpha^{(2)}$ and $\alpha^{(3)}$ are the nonlinear polarizabilities and $E(r)$ is the local field at site r. The polarization of a medium (i.e. dipole moment per unit volume), which is a source of the electromagnetic field in a medium, can be represented in an analogous form with the coefficients $\chi^{(n)}$ called "susceptibilities." When the local field considerably exceeds the applied field, $E^{(0)}$, huge enhancements of nonlinear optical responses occur.

2.2 Surface-Enhanced Optical Responses

Below, we consider the average enhancement for nonlinear optical responses in an arbitrary random system where the local fields experience fluctuations. In later chapters we apply these formulas, or their variations, for specific random systems, such as aggregates of colloidal particles, self-affine thin films, and semicontinuous metal films. Hereafter, we assume that each site of an object possesses a required nonlinear polarizability in addition to the linear one.

For a special case of fractal objects, as mentioned above, the local fields associated with the light-induced eigenmodes can significantly exceed the applied macroscopic field $\mathbf{E}^{(0)}$, leading to an especially large enhancement of optical nonlinearities. For metal composites, this enhancement typically increases toward the infrared part of the spectrum, where resonance quality-factors are significantly larger [11, 53, 54] (see also Chap. 3).

2.2.1 Kerr Optical Nonlinearity

We begin our consideration with $\chi^{(3)}(-\omega;\omega,\omega,-\omega)$, the Kerr-type optical nonlinearity, which is responsible for nonlinear corrections to absorption and refraction. This type of nonlinearity can be used, in particular, for optical switches and optical limiters. The local nonlinear dipole is in this case proportional to $|\mathbf{E}(\mathbf{r})|^2\mathbf{E}(\mathbf{r})$, where $\mathbf{E}(\mathbf{r})$ is the local field at site $\mathbf{r}$. For the resonant optical modes, the local fields exceed the macroscopic (average) field by a resonance quality factor q.

The fields generated by the nonlinear dipoles can also excite the eigenmodes of a surface, resulting in secondary enhancement $\propto \mathbf{E}(\mathbf{r})/\mathbf{E}^{(0)}$. Accordingly, the surface-enhanced Kerr-susceptibility, $\bar{\chi}^{(3)}$, can (with the angular brackets in the following formulas denoting an ensemble-average) be written as [11, 20, 45, 53, 54]

$$\bar{\chi}^{(3)}/p\chi^{(3)} = G_{\mathrm{K}} = \frac{\left\langle \left|\mathbf{E}(\mathbf{r})\right|^2 \left[\mathbf{E}(\mathbf{r})\right]^2 \right\rangle}{\left[\mathbf{E}^{(0)}\right]^4} . \tag{2.1}$$

Here $\chi^{(3)}$ is the initial "seed" susceptibility. It can be associated with some molecules adsorbed on a metal surface – and then $\bar{\chi}^{(3)}$ represents the nonlinear susceptibility of the composite material consisting of the adsorbed nonlinear molecules and a surface providing the enhancement. Alternatively, the seed $\chi^{(3)}$ can be associated with the metal particles themselves. Then, G_K represents the enhancement due to clustering of initially isolated particles into aggregates, with the average volume fraction of the metal given by p. Note that G_K depends on the local-field phases and contains, in general, both real and imaginary parts.

The applied field with the frequency in the visible and near infrared (IR) parts of the spectrum is typically off-resonance for a nearly spherical colloidal particle (e.g. silver) but it efficiently excites the collective modes of an aggregate of the particles – for example, the fractal eigenmodes cover a large frequency interval including the visible and infrared parts of the spectrum [11, 53, 54] (see Chap. 3).

For simplicity, we assume that $\mathbf{E}^{(0)}$ in (2.1) is linearly polarized and can therefore be chosen as real. The formula in (2.1) is to be proven from rigorous first-principle considerations in Subsect. 3.6.4.

2.2.2 Four-Wave Mixing

Four-wave mixing (FWM) is a four-photon process that is determined by the nonlinear susceptibility similar to (2.1): $\chi^{(3)}_{\alpha\beta\gamma\delta}(-\omega_s;\omega_1,\omega_1,-\omega_2)$, where $\omega_s = 2\omega_1 - \omega_2$ is the generated frequency, and ω_1 and ω_2 are the frequencies of the applied waves.

Coherent anti-Stokes Raman scattering (CARS) is an example of FWM. In one elementary act of the CARS process, two ω_1 photons are converted into ω_2 and ω_s photons. Another example is degenerate FWM (DFWM); this process is used for optical phase conjugation (OPC) that can result in the complete removal of optical aberrations [55]. In DFWM, all waves have the same frequency ($\omega_s = \omega_1 = \omega_2$) and differ only in their propagation directions and, in general, in their polarizations. In a typical OPC experiment, two oppositely directed pump beams, with field amplitudes $\mathbf{E}^{(1)}$ and $\mathbf{E}'^{(1)}$, and a probe beam (propagating at some small angle to the pump beams) with amplitude $\mathbf{E}^{(2)}$ result in a generated OPC beam that propagates against the probe beam. Because of the interaction geometry, the wave vectors of the beams satisfy the equation $\mathbf{k}_1 + \mathbf{k}'_1 = \mathbf{k}_2 + \mathbf{k}_s = 0$. Clearly, for two pairs of oppositely directed beams with the same frequency ω, the phase-matching condition (resulting from the conservation of photon momentum) is automatically fulfilled [55].

The nonlinear susceptibility $\chi^{(3)}$ that results in DFWM also leads to the nonlinear refraction and absorption that are associated with the Kerr optical nonlinearity. Note that, as above, the nonlinear susceptibility $\chi^{(3)}$ can be associated with either the metal particles or the molecules adsorbed on the metal surface.

For coherent effects, including the ones discussed in this section, the ensemble averaging is performed for the generated field amplitude (rather than the intensity) – in other words, for the nonlinear polarization of a random medium. The average polarization $P^{(3)}(\omega)$ is proportional to the nonlinear susceptibility, i.e. $P^{(3)}(\omega) \propto \bar{\chi}^{(3)} = p\chi^{(3)}G_{\mathrm{K}}$. The measured signal for coherent processes is proportional to $|\bar{\chi}^{(3)}|^2$. Thus we conclude that the resultant enhancement for degenerate (or nearly degenerate) four-wave mixing can be expressed in terms of enhancement of Kerr susceptibility as follows [54]:

$$G_{\mathrm{FWM}} = \left|G_{\mathrm{K}}\right|^2 = \left|\frac{\left\langle \left|\mathbf{E}(\mathbf{r})\right|^2 \left[\mathbf{E}(\mathbf{r})\right]^2 \right\rangle}{\left[\mathbf{E}^{(0)}\right]^4}\right|^2 . \tag{2.2}$$

Note that it is possible equally to describe a medium optical response in terms of nonlinear currents rather than nonlinear polarizations. These two approaches are completely equivalent (see also Chap. 5).

2.2.3 Raman Scattering

Raman scattering is a linear optical process. For small Stokes shifts, however, the surface-enhanced Raman scattering is proportional to a mean of the fourth power of the local fields [51, 52]. This is because the local Stokes fields get enhanced on a resonating metal surface, along with the field at the fundamental frequency, so that the resulting enhancement factor for Raman scattering is given by

$$G_{\mathrm{RS}} = \frac{\left\langle \left|\mathbf{E}(\mathbf{r})\right|^4 \right\rangle}{\left|\mathbf{E}^{(0)}\right|^4} . \tag{2.3}$$

Note that, in contrast to the enhanced Kerr nonlinearity considered above, G_{RS} is real and the local enhancement is phase insensitive, so that there is no possible destructive interference of signals from different points of a surface.

In Chap. 3, we derive the above formula from first-principle considerations (see Subsect. 3.6.2).

2.2.4 High-Order Harmonic Generation

Under some simplifying conditions, enhancement for second-harmonic generation (SHG) can be written as [11]

$$G_{\mathrm{SHG}} = \left|\left\langle \left[\frac{E_\omega(\mathbf{r})}{E_\omega^{(0)}}\right]^2 \left[\frac{E_{2\omega}(\mathbf{r})}{E_{2\omega}^{(0)}}\right] \right\rangle\right|^2 , \tag{2.4}$$

where $E_{2\omega}^{(0)}$ and $E_{2\omega}(\mathbf{r})$ are the macroscopic (probe) and local *linear* fields at frequency 2ω. If there is no additional surface enhancement at frequency 2ω, then $E_{2\omega}(\mathbf{r}) = E_{2\omega}^{(0)}$.

In Subsect. 4.3.2, we derive a formula similar to (2.4) for the general case.

The formula in (2.4) can easily be generalized to the nth harmonic generation (nHG):

$$G_{n\mathrm{HG}} = \left| \left\langle \left[\frac{E_{\omega}(\mathbf{r})}{E_{\omega}^{(0)}} \right]^{n} \left[\frac{E_{n\omega}(\mathbf{r})}{E_{n\omega}^{(0)}} \right] \right\rangle \right|^{2} . \tag{2.5}$$

We can note that formulas (2.1)–(2.5) are valid for an arbitrary surface. In a special case of random films, because of the extremely large field fluctuations the ensemble-average enhancements are typically much larger than those for nonrandom surfaces. In addition, the fractal and percolation composites provide enhancements in a very large spectral range, including the infrared part of the spectrum, where enhancement is particularly large because of localization of optical excitations and large resonance quality factors [11].

It will be shown below that local enhancements in the hot spots can exceed the ensemble-average enhancement by many orders of magnitude; this opens up a new feasibility for local surface-enhanced nanospectroscopy, as discussed in Chap. 3.

In the following chapters we derive formulas similar to (2.1)–(2.5), using a rigorous approach, and apply them for calculation of enhancement of optical phenomena in various random systems. As mentioned, these surface-enhanced optical effects, briefly outlined here, result from high local fields associated with the resonance optical excitations in random media.

We note that in this book we are focused on the electromagnetic enhancement due to surface plasmons. This mechanism is a nonselective amplifier for optical processes in all molecules as opposed to the chemical enhancement. The latter can be due to the shift of the electronic states of the adsorbate by interaction with the surface or due to new electronic states arising from chemisorption and serving as resonant intermediate states. In contrast to the electromagnetic mechanism, the chemical enhancement can occur only for selected molecules.

3. Small-Particle Fractal Aggregates

Murray Gell-Mann spent a semester staying in Schwinger's house in Cambridge and loved to say afterwards that he had searched everywhere for the Feynman diagrams. He had not found any, but one room had been locked ...

James Gleick, *Genius*

3.1 Introduction

Random fractal clusters are complex systems built from simple elementary blocks, e.g. particles, that are called monomers. It is important to emphasize that the rich and complicated properties of fractal clusters are determined by their global geometrical structure rather than by the structure of each monomer. In the formulation of a typical problem in optics of fractal clusters, the properties of monomers and the laws of physics for their interaction with the incident field and with each other are known, while the properties of a cluster as a whole must be found.

In Fig. 3.1, you see a picture of a typical fractal aggregate of silver colloidal particles (obtained via an electron-microscope). The fractal dimension of these aggregates is $D \approx 1.78$. Using the well known model of cluster-cluster aggregation, colloidal aggregates can be readily simulated numerically [7]. Note that voids are present at all scales, from the minimum (about the size of a single particle) to the maximum (about the size of the whole cluster); this is an indication of the statistical self-similarity of a fractal cluster. The size of an individual particle is $\sim$ 10nm, whereas the size of the whole cluster is $\sim$ 1μm.

The process of aggregation, resulting in clusters similar to the one shown in Fig. 3.1, can be described as follows. A large number of initially isolated and randomly distributed nanoparticles execute random walks in the solution. Encounters with other nanoparticles result in their sticking together, first to form small groups, and then, in the course of the random walking, to aggregate into larger formations, and so on. Cluster-cluster aggregates (CCAs), with the fractal dimension $D \approx 1.78$, are thereby eventually formed.

Early advances in the study of *nonresonant* scattering of light and X-rays by fractal clusters were accomplished by Bale and Schmidt [56], Berry

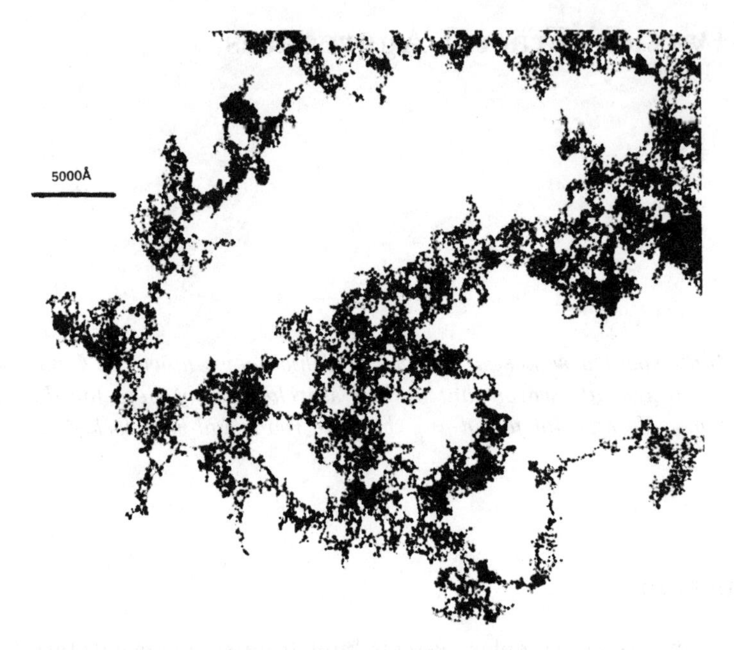

Fig. 3.1. Electron micrograph of a fractal colloid aggregate

and Percival [57], and Martin and Hurd [58]. Since then the theory of non-resonant scattering by fractals has been developed in great detail and applied to various fractal objects [59–65].

The foundations of the *resonance* optics of fractals were built in papers by Stockman, Markel, the author of this book, and their co-authors [45, 46, 66, 67]. Here in the chapter introduction we briefly mention some basic ideas and results of the resonance optics of fractals, and in the following sections of this chapter we will consider fractal optical properties in detail.

The light-induced plasmon oscillations in different metal particles of a fractal result in their polarization. In contrast to conventional (non-fractal) three-dimensional systems, in fractals the interactions between the resonance dipoles induced in particles are not of a long-range character. This results in spatial localization of the optical excitations at various random locations in a fractal cluster [11, 45–49]. The spectrum of these optical eigenmodes exhibits strong inhomogeneous broadening as a result of a large variety in the local configurations in a random fractal. The basic optical properties of fractals can be understood from the following qualitative considerations.

In a fractal cluster, there always exists a high probability of finding some particles in close proximity to any given particle, something that accounts for strong interparticle interactions. More precisely, in fractals, according to (1.2), the pair correlation function $g \propto r^{D-d}$ becomes large at small r (where D is the fractal dimension and d is the dimension of the embedding space; $D < d$), and this implies that there are always close neighbors for any particle. On the other hand, according to (1.1), the number of particles in a fractal cluster is given by $N = (R_c/R_0)^D$ (where R_c and R_0 are the size of a cluster and typical separation between neighbor particles); therefore, the volume fraction p filled by the particles is very small, such that $p \propto (R_c/R_0)^{D-d} = N^{1-d/D} \rightarrow 0$, i.e. it tends to zero with increasing size (or, equivalently, with increasing number of particles). Thus, objects with fractal morphology possess an unusual combination of properties: despite the fact that the volume fraction filled by particles in a fractal is very small (as, say, in a low-density gas), strong interactions exist between the particles (as, for example, in a condensed matter) [45]. These strong interactions between the neighboring particles result in forming the collective modes that may strongly differ in resonance frequencies, because of different local geometries where the modes are localized in a cluster. This eventually leads to the inhomogeneously broadened spectrum covering a broad spectral range [45, 53]. Note that, in the picture described, an important fact is that optical excitations are localized in small parts of a fractal, so that the mode resonance frequencies depend on the *local* configurations, which vary a lot in a random fractal.

Localization of optical excitations in fractals leads to a patchwork-like distribution of the local fields associated with hot (strong field) and cold (weak field) zones [47–50, 72]. This in turn results in large spatial fluctuations of the local fields in a fractal and, eventually, in giant enhancements of optical effects [11, 54, 68–74].

It is interesting to note that a fractal itself can be considered (when it is mathematically defined) as a fluctuation, and this is because the fractal's average density is asymptotically zero. The fluctuative nature of fractals is fully manifested in the resonance optical properties of fractal clusters, as described below in this chapter.

Note that, as shown by Stockman and his colleagues [48, 49], a pattern of localization of optical modes in fractals is complicated and can be called inhomogeneous. At any given frequency, individual eigenmodes are dramatically different from each other and their sizes (their coherent radii) vary in a wide range, from the size of an individual particle to the size of a whole cluster. (In the vicinity of the plasmon resonance of individual particles, even chaotic behavior of the eigenmodes in fractals can be found [49].) However, even delocalized modes typically consist of two or more very sharp peaks that are topologically disconnected, i.e. located at relatively large distances from each other. In any case, the electromagnetic energy is mostly concentrated in these peaks, which can belong to different modes or, sometimes, to the same mode. Thus despite the complex inhomogeneous pattern of localization in fractals, there are always very sharp peaks, where the local fields are high. These hot spots eventually provide significant enhancement for a number of optical processes, especially the nonlinear ones that are proportional to the local fields raised to a power greater than one.

Localization of optical excitations in fractals (first suggested in [45]) and strong enhancement of the local fields and optical nonlinearities have been intensively studied during the last decade, both theoretically and experimentally [11, 45–50, 52–54, 66–79]. In experimental studies, a leading role was played by Safonov and his coworkers [73, 74, 78, 79], who performed a number of pioneering experiments on the resonance optics of fractals. For example, in accordance with theoretical predictions [45, 46], a 10^6-fold enhancement of degenerate four-wave mixing was observed when silver colloidal particles aggregated into fractal clusters [73, 78]. For fractal aggregates, wavelength- and polarization-selective spectral holes were detected in the absorption spectra, after irradiation by intense laser pulses [79]. Also, strongly enhanced Kerr-type nonlinearities in absorption and refraction were observed [74]. In the following sections, we consider all these fractal-enhanced nonlinearities in more detail.

Beforehand, though, we consider theoretical approaches to describing the resonance optical properties of an arbitrary ensemble of particles and then apply them to fractal aggregates.

3.2 Basic Theoretical Approaches

In this section we consider a cluster of N monomers located at the points $\mathbf{r}_i$, $i = 1, ..., N$ and interacting with an incident plane monochromatic wave of the form

$$\mathbf{E}_{\text{inc}}(\mathbf{r}, t) = \mathbf{E}^{(0)} \exp(\mathrm{i}\omega t - \mathrm{i}\mathbf{k} \cdot \mathbf{r}) \ . \tag{3.1}$$

The factor $\exp(i\omega t)$ is common for all time-varying fields and will be omitted below.

3.2.1 Coupled Dipoles

In this model, each monomer in a cluster is considered to be a point dipole with polarizability α_0 located at point $\mathbf{r}_i$ (at the center of the respective spherical monomer). The dipole moment of the ith monomer, $\mathbf{d}_i$, is proportional to the local field at the point $\mathbf{r}_i$ that is a superposition of the incident field and all the secondary fields scattered by other dipoles. Therefore the dipole moments of the monomers are coupled to the incident field and to each other as described by the coupled-dipole equation (CDE):

$$\mathbf{d}_i = \alpha_0 \left[\mathbf{E}_{\text{inc}}(\mathbf{r}_i) + \sum_{j \neq i}^{N} \hat{G}(\mathbf{r}_i - \mathbf{r}_j)\mathbf{d}_j \right] \ . \tag{3.2}$$

Here the term $\hat{G}(\mathbf{r}_i - \mathbf{r}_j)\mathbf{d}_j$ gives the dipole radiation field created by the dipole $\mathbf{d}_j$ at the point $\mathbf{r}_i$, and $\hat{G}(\mathbf{r})$ is the regular part of the free space dyadic Green's function:

$$G_{\alpha\beta}(\mathbf{r}) = k^3 \left[A(kr)\delta_{\alpha\beta} + B(kr)r_\alpha r_\beta/r^2\right] , \tag{3.3}$$

$$A(x) = [x^{-1} + \mathrm{i}x^{-2} - x^{-3}] \exp(\mathrm{i}x) , \tag{3.4}$$

$$B(x) = [-x^{-1} - 3\mathrm{i}x^{-2} + 3x^{-3}] \exp(\mathrm{i}x) , \tag{3.5}$$

where $\hat{G}\mathbf{d} = G_{\alpha\beta}d_\beta$. The Greek indices stand for the Cartesian components of vectors and summation over repeated indices is implied. The operator $\hat{G}$ is completely symmetrical: $\hat{G}(\mathbf{r}) = \hat{G}(-\mathbf{r})$, $G_{\alpha\beta}(\mathbf{r}) = G_{\beta\alpha}(\mathbf{r})$. While the monomer size is assumed to be small compared with the wavelength, the overall cluster size is, in general, arbitrary. That is why the near-, intermediate- and far-zone terms ($\propto r^{-3}$, r^{-2} and r^{-1} respectively) are included in formulas (3.4) and (3.5).

The CDE is a system of $3N$ linear equations that can be solved to find the dipole moments $\mathbf{d}_i$. The scattering amplitude is expressed through the dipole moments as

$$\mathbf{f}(\mathbf{k}') = k^2 \sum_{i=1}^{N} \left[\mathbf{d}_\mathrm{i} - (\mathbf{d}_i \cdot \mathbf{k}')\mathbf{k}'/k^2\right] \exp(-i\mathbf{k}' \cdot \mathbf{r}_i) , \tag{3.6}$$

where $\mathbf{k}'$ is the scattered wavevector (which gives the direction of scattering) and $|\mathbf{k}'| = |\mathbf{k}|$. The cross-sections of extinction, scattering, and absorption can be found from the optical theorem:

$$\sigma_\mathrm{e} = \frac{4\pi}{k} \frac{\mathrm{Im}[\mathbf{f}(\mathbf{k}) \cdot \mathbf{E}^{(0)*}]}{|\mathbf{E}^{(0)}|^2} = \frac{4\pi k}{|\mathbf{E}^{(0)}|^2} \mathrm{Im} \sum_{i=1}^{N} \mathbf{d}_i \cdot \mathbf{E}^*_\mathrm{inc}(\mathbf{r}_i) , \tag{3.7}$$

$$\frac{\mathrm{d}\sigma_s}{\mathrm{d}\Omega} = |\mathbf{f}(\mathbf{k})|^2 \ ; \ \sigma_s = \int |\mathbf{f}(\mathbf{k})|^2 \mathrm{d}\Omega , \tag{3.8}$$

$$\sigma_a = \sigma_\mathrm{e} - \sigma_s . \tag{3.9}$$

It can be shown [80] by direct integration in (3.8) over the spatial angles Ω, and with the use of (3.2)–(3.6), that the integral scattering and absorption cross-sections can be expressed through the dipole moments as

$$\sigma_s = \frac{4\pi k}{|\mathbf{E}^{(0)}|^2} \sum_{i=1}^{N} \left\{\mathrm{Im}\left[\mathbf{d}_i \cdot \mathbf{E}^*_\mathrm{inc}(\mathbf{r}_i)\right] - y_a|\mathbf{d}_i|^2\right\} , \tag{3.10}$$

$$\sigma_a = \frac{4\pi k}{|\mathbf{E}^{(0)}|^2} y_a \sum_{i=1}^{N} |\mathbf{d}_i|^2 , \tag{3.11}$$

$$y_a = -\mathrm{Im}\left(\frac{1}{\alpha_0}\right) - \frac{2k^3}{3} \geq 0 . \tag{3.12}$$

Note that the constant y_a is non-negatively defined [80] for any physically reasonable α_0. The ratio $3y_a/2k^3$ characterizes the relative strength of absorption by a single isolated monomer.

3.2.2 Polarizability of a Monomer

In order to solve the CDE (3.2), it is necessary to specify not only the position of monomers but also the polarizability α_0, which plays the role of the coupling constant. We discuss here methods for defining α_0.

The CDE was originally proposed by Purcell and Pennypacker [81] for numerical analysis of scattering and absorption of light by non-spherical dielectric particles. In the formulation by Purcell and Pennypacker, an arbitrary dielectric object is represented by an array of point dipoles placed on a cubic lattice and restricted by the surface of the object. The polarizability of an elementary dipole α_0 is defined by the Clausius-Mossotti relation. Note that it is also equal to the polarizability of a little sphere of such a radius that the total volume of all the spheres is equal to the total volume of the object under investigation. To satisfy this equality, the following relation between the lattice period, a, and the radius of a sphere, $R_{\rm m}$, must hold: $a^3 = (4\pi/3)R_{\rm m}^3$. Note that two spheres of radius $R_{\rm m}$ placed in neighboring sites of a lattice with the period a would geometrically intersect because $a/R_{\rm m} = (4\pi/3)^{1/3} \approx 1.612 < 2$. The polarizability of an elementary dipole in the Purcell-Pennypacker model is given by the Lorentz-Lorenz formula (called also the Clausius-Mossotti relation) [81] with the correction for radiative reaction introduced later by Draine [82]:

$$\alpha_0 = \frac{\alpha_{\rm LL}}{1 - {\rm i}(2k^3/3)\alpha_{\rm LL}} , \tag{3.13}$$

$$\alpha_{\rm LL} = R_m^3 \frac{\epsilon - 1}{\epsilon + 2} , \tag{3.14}$$

where ϵ is the dielectric constant of the material and $\alpha_{\rm LL}$ is the Lorentz-Lorenz polarizability without the radiation correction. Note that polarizability written in the form of (3.13) provides a positive value of y_a in (3.12). More sophisticated formulas for α_0, containing several first terms in the expansion of the polarizability with respect to the parameter $kR_{\rm m}$, can be found in papers by Lakhtakia [83], Draine and Goodman [84], and in the references therein.

In the case of fractal clusters, the situation is different from the one described above. Instead of the imaginary spheres, we have real touching spherical monomers that form a cluster. Of course, these monomers cannot intersect geometrically. Let the radius of the real monomers that can be seen experimentally be $R_{\rm exp}$, and the distance between two neighboring monomers be $a_{\rm exp}$. For touching spheres $a_{\rm exp}/R_{\rm exp} = 2$. However, it has been shown both theoretically [85, 86] and experimentally [87] that the CDE (3.2) with $a_{\rm exp}/R_{\rm exp} = 2$, α_0 given by (3.13) and (3.14), $a = a_{\rm exp}$ and $R_{\rm m} = R_{\rm exp}$ yields incorrect results. The reason for this is that the dipole field generated by one monomer is not homogeneous inside the adjacent particle; it is much stronger near the point where the monomers touch than in the center of the neighboring monomer. (Therefore, strictly speaking, when the field is not homogeneous inside a dielectric sphere, one should not restrict consideration just to dipole moments; see, however, the discussion below.) Effectively, by replacing two touching spheres with two point dipoles located in their centers, we underestimate the actual strength of their interaction.

To compensate the above fact, a model for a fractal cluster was introduced [53] in which neighboring spheres were allowed to intersect geometrically. The radii of these spheres, as well as the distance between two neighboring monomers, are chosen to be different from the experimental ones: $R_{\rm m} \neq R_{\rm exp}$, $a \neq a_{\rm exp}$; but it is required that the ratio $a/R_{\rm m}$ is equal to $(4\pi/3)^{1/3} \approx 1.612$, the same as in the Purcell and Pennypacker model [81]. Note that a close value for the above ratio, $a/R_{\rm m} \approx 1.688$, was obtained from the condition that an infinite linear chain of polarizable spherules has the correct depolarization coefficient [80]. The second equation for $R_{\rm m}$ and a can be obtained from the optically important condition that the model cluster has the same fractal dimension, radius of gyration and total volume as the experimental one. The two equations can be satisfied simultaneously for nontrivial fractal clusters (with $D < d = 3$) and lead to

$$R_{\rm m} = R_{\rm exp}(\pi/6)^{D/[3(3-D)]} \; ; \quad N = N_{\rm exp}(6/\pi)^{D/(3-D)} , \tag{3.15}$$

where $N_{\rm exp}$ and N are the number of monomers in the original and in the model cluster, respectively. The invariance of the radius of gyration, $R_{\rm c}$, follows from (1.1) and (3.15).

The above model was shown to yield results that are in very good agreement with experimental spectra of fractal clusters [53]. We need also to emphasize that the model of the effective intersecting particles allows us phenomenologically to take into account the stronger depolarization factors for touching particles. Typically, we need to involve multipolar corrections for calculating these depolarization factors. In contrast, the described model, although inexact, allows us to obtain the correct optical characteristics while remaining within the dipolar approximation - which is, of course, crucial for a complex random system, such as a fractal cluster consisting of thousands of particles.

3.2.3 Eigenmode Expansion

The CDE (3.2) takes an elegant form when written in matrix notation. We introduce a linear complex vector space C^{3N} and $3N$-dimensional vectors of the dipole moments and incident fields according to $|{\rm d}\rangle = (\mathbf{d}_1, \mathbf{d}_2, ..., \mathbf{d}_N)$ and $|{\rm E}\rangle = (\mathbf{E}_{\rm inc}(\mathbf{r}_1), \mathbf{E}_{\rm inc}(\mathbf{r}_2), ..., \mathbf{E}_{\rm inc}(\mathbf{r}_N))$. We also define an orthonormal basis $|i\alpha\rangle$ in C^{3N}, such that the Cartesian components of the dipole moments are expressed in this basis as $d_{i\alpha} = \langle i\alpha|{\rm d}\rangle$. The linear operator V acts on the vector of dipole moments $|{\rm d}\rangle$ according to the rule: $\langle i\alpha|{\rm V}|{\rm d}\rangle = \sum_{\beta j} G_{\alpha\beta}(\mathbf{r}_i - \mathbf{r}_j)d_{j\beta}$. Then (3.2) can be written as:

$$|\mathrm{d}\rangle = \alpha_0 \left(|\mathrm{E}\rangle + \mathrm{V}|\mathrm{d}\rangle\right) \tag{3.16}$$

and the expressions for the optical cross-sections acquire the form:

$$\sigma_{\mathrm{e}} = \frac{4\pi k}{|\mathbf{E}^{(0)}|^2} \mathrm{Im}\langle \mathrm{E}|\mathrm{d}\rangle \ , \tag{3.17}$$

$$\sigma_a = \frac{4\pi k y_a}{|\mathbf{E}^{(0)}|^2} \langle \mathrm{d}|\mathrm{d}\rangle \ . \tag{3.18}$$

The interaction matrix V is complex and symmetrical and, consequently, non-Hermitian. Therefore, its eigenvectors are not, in general, orthogonal in C^{3N}. However, it can be shown [80, 88] that the eigenvectors are linearly independent, and therefore form a basis, if V is not degenerate or if the nature of its degeneracy is "geometrical" (due to a certain symmetry of the cluster under consideration) rather than "random". In most practical cases, we can define a complete basis of eigenvectors, $|\mathrm{n}\rangle$, corresponding to the eigennumbers v_n:

$$\mathrm{V}|\mathrm{n}\rangle = v_n |\mathrm{n}\rangle \ . \tag{3.19}$$

For complex symmetrical matrices, the orthogonality rule ($\langle \mathrm{m}|\mathrm{n}\rangle = \delta_{mn}$) is replaced by [80, 88]

$$\langle \bar{\mathrm{m}}|\mathrm{n}\rangle = 0 \ \text{ if } m \neq n \ , \tag{3.20}$$

where the bar denotes complex conjugation of all elements of a vector. Thus, $|\mathrm{n}\rangle$ denotes a column vector, $\langle \mathrm{n}|$ denotes a row vector with the complex conjugated elements, and $\langle \bar{\mathrm{n}}|$ denotes a row vector with exactly the same elements as $|\mathrm{n}\rangle$. We adopt the usual normalization of the eigenvectors, $\langle \mathrm{n}|\mathrm{n}\rangle =$ 1, but $\langle \bar{\mathrm{n}}|\mathrm{n}\rangle$ is not equal to unity and can be, in general, complex.

The representation of the unity operator in the basis $|\mathrm{n}\rangle$ is

$$\mathrm{I} = \sum_{n=1}^{3N} \frac{|\mathrm{n}\rangle\langle \bar{\mathrm{n}}|}{\langle \bar{\mathrm{n}}|\mathrm{n}\rangle} \ . \tag{3.21}$$

The eigenvector expansion for the solution to the CDE, $|\mathrm{d}\rangle$, can be easily obtained with the use of the above equality [53]:

$$|\mathrm{d}\rangle = \sum_{n=1}^{3N} \frac{|\mathrm{n}\rangle\langle \bar{\mathrm{n}}|\mathrm{E}\rangle}{\langle \bar{\mathrm{n}}|\mathrm{n}\rangle(1/\alpha - v_n)} \ . \tag{3.22}$$

In the $|i\alpha\rangle$ basis, the solution acquires the form:

$$d_{i,\alpha} = \sum_{n,j} \frac{\langle i\alpha|n\rangle\langle \bar{n}|j\beta\rangle E_{j,\beta}}{\left[\sum_{i'}\langle \bar{n}|i'\alpha'\rangle\langle i'\alpha'|n\rangle\right]} \frac{1}{1/\alpha_0 + v_n} . \tag{3.23}$$

The above formulas indicate that for an arbitrary collection of N interacting particles, there are $3N$ eigenmodes, with their resonant eigenfrequencies defined by $\mathrm{Re}(1/\alpha_0) + v_n = 0$. The weight with which a mode contributes to the resultant optical response depends on the scalar product $\langle \bar{n}|E\rangle$ and thus on the symmetry properties of the eigenvectors $|n\rangle$.

Equations (3.17), (3.18) and (3.22) give the general solutions to the CDE and dependence of the optical cross-sections on the polarizability α_0.

The following two exact properties [80, 88] of the eigenvalues of the CDE can be useful for assessment of accuracy of different numerical methods for diagonalization of V:

$$\sum_{n=1}^{3N} v_n = 0 \; ; \quad -2k^3/3 \le \mathrm{Im} v_n \le (3N-1)2k^3/3 \; . \tag{3.24}$$

3.2.4 Quasi-Static Approximation

In this subsection we restrict our consideration to the quasi-static limit, where the characteristic system size L is assumed to be much smaller than the wavelength $\lambda = 2\pi c/\omega$. (We note, however, that, as shown in [66], for fractals with $D < 2$, the quasi-static approximation also describes well the resonance optical properties of clusters of arbitrary sizes, even with $L \gg \lambda$.)

In the quasi-static approximation, we leave only the near-field term in the expression for the Green function and the factor $\exp(\mathrm{i}\mathbf{k}\cdot\mathbf{r}_i)$ can be replaced by unity. In addition, since the time dependence, $\exp(-\mathrm{i}\omega t)$, is the same for all time-varying fields, the whole exponential factor $\exp(\mathrm{i}\omega t - \mathrm{i}\mathbf{k}\cdot\mathbf{r})$ can be omitted in the quasi-static approximation. After that, the coupled-dipole equations (CDE) (3.2)–(3.5) for the induced dipoles acquire the following form [11, 45, 46]:

$$d_{i,\alpha} = \alpha_0 \left(E_\alpha^{(0)} + \sum_{j\neq i} W_{ij,\alpha\beta} d_{j,\beta} \right) , \tag{3.25}$$

$$W_{ij,\alpha\beta} = \langle i\alpha|W|j\beta\rangle = \left(3r_{ij,\alpha}r_{ij,\beta} - \delta_{\alpha\beta}r_{ij}^2\right)/r_{ij}^5, \tag{3.26}$$

where $W_{ij,\alpha\beta}$ is the quasi-static interaction operator between two dipoles, $\mathbf{r}_i$ is the radius-vector of the ith monomer and $\mathbf{r}_{ij} = \mathbf{r}_i - \mathbf{r}_j$. The Greek indices, as above, denote Cartesian components of vectors and should not be confused with the polarizability α_0.

The linear polarizability of an elementary dipole representing a spherical monomer α_0 is given by the Lorentz-Lorenz formula:

$$\alpha_0 = R_\mathrm{m}^3[(\epsilon - \epsilon_\mathrm{h})/(\epsilon + 2\epsilon_\mathrm{h})], \tag{3.27}$$

where, as above, $\epsilon = \epsilon' + \mathrm{i}\epsilon''$ is the bulk dielectric permittivity of the particles and ϵ_h in the dielectric constant of the host material. We also neglected the radiation term in the more general formula (3.13) since it is typically small in metal particles, and we hereafter set $\alpha_0 = \alpha_\mathrm{LL}$. The radius R_m is chosen according to the model of intersecting effective spheres described in Sect. 3.2.2.

Recall that, for metal particles, the dielectric function is well described by the Drude formula (see also (1.16)–(1.18))

$$\epsilon = \epsilon_0 - \frac{\omega_p^2}{\omega(\omega + \mathrm{i}\Gamma)}, \tag{3.28}$$

where ϵ_0 includes a contribution to the dielectric constant associated with interband transitions in a bulk material, ω_p is the plasma frequency and $\Gamma = 1/\tau \equiv \omega_\tau$ is the relaxation constant, where τ is the plasmon relaxation time.

Since $W_{ij,\alpha\beta}$ is independent of the frequency ω in the quasi-static approximation, the spectral dependence of solutions to (3.25) enters only through $\alpha_0(\omega)$. For convenience, we introduce the variable

$$Z(\omega) \equiv 1/\alpha_0(\omega) = -[X(\omega) + \mathrm{i}\delta(\omega)]. \tag{3.29}$$

Using (3.27), we obtain

$$X \equiv -\mathrm{Re}[\alpha_0^{-1}] = -R_\mathrm{m}^{-3}[1 + 3\epsilon_\mathrm{h}(\epsilon' - \epsilon_\mathrm{h})/|\epsilon - \epsilon_\mathrm{h}|^2], \tag{3.30}$$

$$\delta \equiv -\mathrm{Im}[\alpha_0^{-1}] = 3R_\mathrm{m}^{-3}\epsilon_\mathrm{h}\epsilon''/|\epsilon - \epsilon_\mathrm{h}|^2. \tag{3.31}$$

The variable X indicates the proximity of ω to the resonance of an individual particle, occurring for a spherical particle at $\epsilon' = -2\epsilon_\mathrm{h}$, and it plays the role of a frequency parameter; δ characterizes dielectric losses. The resonance quality factor is proportional to δ^{-1}. At the resonance of a spherical particle when $\epsilon' = -2\epsilon_\mathrm{h}$, we have $(R_\mathrm{m}^3\delta)^{-1} = (3/2)|\epsilon'/\epsilon''|$. However, for collective resonances of an ensemble of particles occurring at $|\epsilon'| \gg \epsilon_\mathrm{h}$ (see below), we have $(R_\mathrm{m}^3\delta)^{-1} \approx |\epsilon|^2/(3\epsilon_\mathrm{h}\epsilon'')$, which increases with the wavelength. One can find $X(\lambda)$ and $\delta(\lambda)$ for any material using theoretical or experimental data for $\epsilon(\lambda)$ and formulas (3.30)–(3.31).

In Figs. 3.2a and 3.2b, plots for X and δ for silver are shown as functions of wavelength λ for two different host media, a vacuum and water. The tabulated data [89] for ϵ were used. The plots are shown in the units in which $(4\pi/3)R_\mathrm{m}^3 = a^3 = 1$ (see Subsect. 3.2.2). As seen in the figure, X changes significantly from 400 nm to 800 nm and then, for longer wavelengths, changes only slightly, so that $X \approx X_0 = -4\pi/3$. The modes with $X \approx X_0$ are referred to as "zero-modes" [11]. The relaxation constant δ is small in the visible spectral range and decreases toward the infrared.

We can write (3.25) in matrix form. Similarly as in the section above, we introduce a $3N$-dimensional vector space R^{3N} and an orthonormal basis $|i\alpha\rangle$. The $3N$-dimensional vector of dipole moments is denoted by $|d\rangle$, and of the incident field by $|E_\mathrm{inc}\rangle$. The Cartesian components of 3-dimensional vectors $\mathbf{d}_i$ and $\mathbf{E}_\mathrm{inc}$ are given by $\langle i\alpha|d\rangle = d_{i,\alpha}$ and $\langle i\alpha|E_\mathrm{inc}\rangle = E_\alpha^{(0)}$. The last equality follows from the assumption that the incident field is uniform

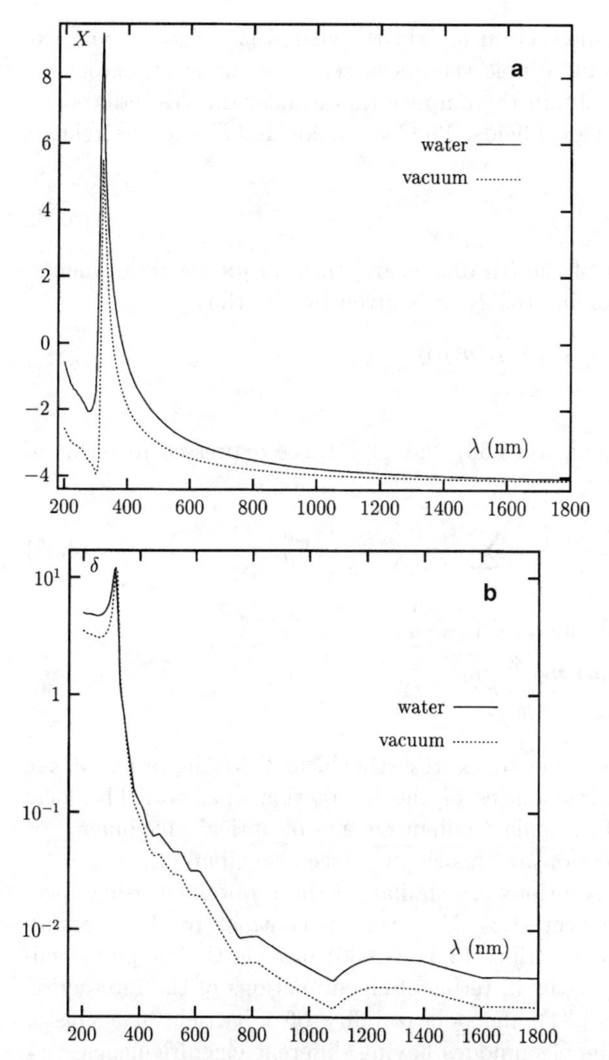

Fig. 3.2. Spectral variable X (**a**) and loss parameter δ (**b**) versus wavelength for silver particles in vacuum and water

throughout the film. The matrix elements of the interaction operator are defined by $\langle i\alpha|\hat{W}|j\beta\rangle = W_{ij,\alpha\beta}$. Then (3.25) can be written as:

$$[Z(\omega) - \hat{W}]|d\rangle = |E_{\text{inc}}\rangle \; . \tag{3.32}$$

The Hermitian interaction operator $\hat{W}$ in (3.32) is real and symmetrical, as can be easily seen from the expression (3.26) for its matrix elements. (Recall that, beyond the quasi-static approximations, the interaction operator is also symmetrical – but it is complex and non-Hermitian.)

By diagonalizing the interaction matrix $\hat{W}$ with $\hat{W}|n\rangle = w_n|n\rangle$ and expanding the $3N$-dimensional dipole vectors in terms of the eigenvectors $|n\rangle$ (as $|d\rangle = \sum_n C_n|n\rangle$), we obtain the amplitudes of linear dipoles induced by the incident wave and the local fields. The local fields and dipoles are related as

$$E_{i,\alpha} = \alpha_0^{-1} d_{i,\alpha} = \alpha_0^{-1} \alpha_{i,\alpha\beta} E_\beta^{(0)}. \tag{3.33}$$

The polarizability tensor of the ith dipole, $\hat{\alpha}_i(\omega)$, with its matrix elements, $\alpha_{i,\alpha\beta}$, can be found by solving (3.32). It is given by [46] thus:

$$\alpha_{i,\alpha\beta} \equiv \alpha_{i,\alpha\beta}(\omega) = \sum_{j,n} \frac{\langle i\alpha|n\rangle\langle n|j\beta\rangle}{Z(\omega) - w_n}. \tag{3.34}$$

The local dipoles, according to (3.33) and (3.34), are expressed in terms of the eigenmodes as follows:

$$d_{i,\alpha} = \sum_n \frac{\langle i\alpha|n\rangle\langle n|E_{\text{inc}}\rangle}{Z(\omega) - w_n} = \sum_{n,j} \frac{\langle i\alpha|n\rangle\langle n|j\beta\rangle}{Z(\omega) - w_n} E_\beta^{(0)}, \tag{3.35}$$

so that the expression for the local field is

$$E_{i,\alpha} = \alpha_0^{-1} \sum_{j,n} \frac{\langle i\alpha|n\rangle\langle n|j\beta\rangle}{Z(\omega) - w_n} E_\beta^{(0)}. \tag{3.36}$$

Equation (3.36) allows one to express the local fields in terms of the eigenfunctions and eigenfrequencies of the interaction operator. The local fields can then be used to calculate enhancements of optical phenomena, as shown in the following sections of this chapter (see also Chap. 2).

Note that the above solutions are similar to those obtained using spectral representations (see Sect. 1.3). However, in contrast to the spectral-representation theory, the solutions (3.34)–(3.36) provide the recipe for calculation of the mode strengths in terms of eigenfunctions of the interaction operator. According to (3.34), the polarizability of a particle in a cluster is given by the sum of the eigenmodes having different eigenfrequencies w_n and contributing with a weight given by the product of the corresponding eigenfunctions.

As follows from (3.34) and (3.39), the conventional Kramers-Kronig formula is valid in terms of X:

$$\text{Re}[\alpha_{i,\alpha\beta}] = \frac{1}{\pi} P \int_{-\infty}^{\infty} \frac{\text{Im}[\alpha_{i,\alpha,\beta}(X')]\text{d}X'}{X' - X},$$

where P denotes the principal value of the integral. Since eigenvalues w_n are real and the decay constant δ is positive, the following exact relations also hold [46]:

$$\frac{1}{\pi} \int_{-\infty}^{\infty} \text{Im}[\alpha_{i,\alpha,\beta}(X)]\text{d}X = \delta_{\alpha,\beta}, \quad P \int_{-\infty}^{\infty} \text{Re}[\alpha_{i,\alpha,\beta}(X)]\text{d}X = 0 \tag{3.37}$$

and

$$P\int_{-\infty}^{\infty} X\mathrm{Im}[\alpha(X)]\mathrm{d}X = 0. \tag{3.38}$$

Note that solution (3.35) to the coupled-dipole equation and the above sum rules are of general character, so that they are valid for an arbitrary cluster, fractal or not.

The average cluster polarizability is given by

$$\alpha = \alpha(X) = (1/3N)\sum_i \mathrm{Tr}[\alpha_{\alpha\beta}^{(i)}]. \tag{3.39}$$

The extinction cross-section σ_e in the quasi-static approximation can be expressed through the imaginary part of the polarizability as

$$\sigma_\mathrm{e} = 4\pi k N \mathrm{Im}[\alpha(X)]. \tag{3.40}$$

For small clusters, scattering can be neglected, so that the extinction cross-section is approximately equal to the absorption cross-section.

3.3 Absorption Spectra of Fractal Aggregates

In this section we discuss the results of numerical and experimental studies of the extinction (absorption) spectra of fractal aggregates of nanoparticles.

To simulate silver colloid aggregates, the cluster-cluster aggregation model can be used [7]. The cluster-cluster aggregates (CCAs) have fractal dimension, structure and aggregation pattern very similar to those observed in experiment. The CCA can be built on a cubic lattice with periodic boundary conditions using a well known algorithm [7].

In Fig. 3.3a, the absorption spectrum, $\mathrm{Im}[\alpha(X)]$, as a function of frequency parameter X is shown for CCAs (units of $a = 1$ are used, where a is the lattice period) [53]. The simulations were performed for clusters consisting of $N =$ 500, 1 000 and 10 000 particles each. Note that the spectrum shows a strong inhomogeneous broadening in CCAs; the spectrum width is much larger than the resonance width for an individual monomer δ (in these simulations $\delta = 0.1$ was used).

It is interesting to compare the absorption spectrum for fractal and non-fractal random media. We consider two nonfractal ensembles of particles: a random gas of particles (RGP) and a close-packed sphere of particles (CPSP). While RGP is a very dilute system of particles randomly distributed in space, CPSP represents a dense (but still random) system of particles. In both cases $D = d = 3$ and the correlation function $g(r)$ is constant. The particles were assumed to be hard spheres. To provide a better comparison with CCA, the RGP was generated in a spherical volume that would be occupied by a CCA with the same number of particles; this means that particles in CCA and RGP fill the same volume fraction p (p was small, with $p \approx 0.05$ for $N = 500$.)

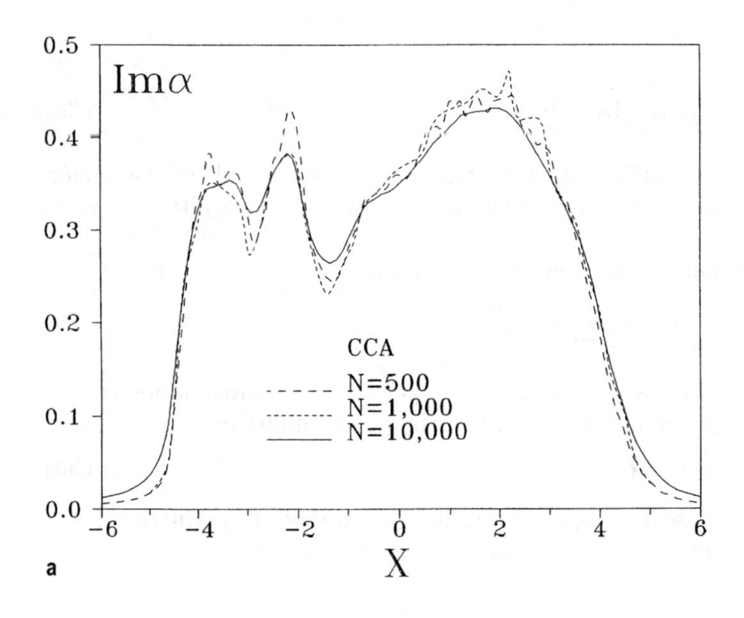

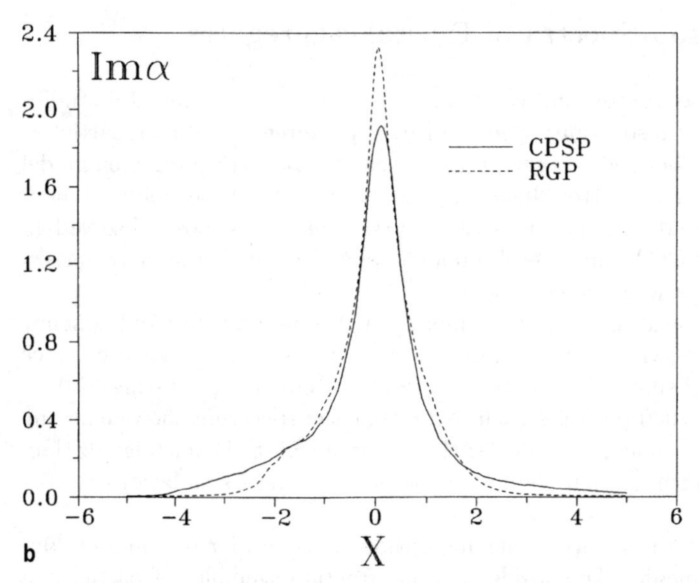

Fig. 3.3. Calculated absorption spectra Im$[\alpha(X)]$ for (**a**) cluster-cluster aggregates (CCAs) containing different number of particles (N = 500, 1 000 and 10 000); and for (**b**) non-fractal 500-particle aggregates of two types: a close-packed sphere of particles (CPSP) and a random gas of particles (RGP)

In contrast, a fairly dense packing of spherical particles, with $p \approx 0.44$, was used for the CPSP.

In Fig. 3.3b, the absorption by a random, nonfractal ensemble of particles, CPSP and RGP, is shown. One can see that the spectral width is in this case much smaller than for fractal cluster-cluster aggregates.

Thus, the dipole-dipole interactions in fractals, in contrast to nonfractal composites (either sparse, like RGP, or compact, like CPSP), result in significantly larger inhomogeneous broadening. (In terms of the optical wavelength, the eigenmodes of silver CCAs, for example, span the visible and infrared parts of the spectrum, while modes in non-fractal silver CPSP and RGP are confined to a narrow range between approximately 350 nm and 450 nm.) This results from the fact that for fractals the dipole-dipole interactions are not long-range, and optical excitations are localized in small areas of a fractal aggregate. As mentioned, these areas have very different local structures and, accordingly, they resonate at different frequencies. In contrast, in compact nonfractal aggregates (with $D = d = 3$), the optical excitations (known also as dipolar collective modes) are delocalized over the whole sample, and their resonance frequencies lie in a relatively narrow spectral interval.

The strong absorption by aggregated particles in the long-wavelength part of the spectrum can be related to the well known enhanced infrared absorption (see, for example, [24]) generally attributed to the clustering of particles. In this part of the spectrum $X \approx X_0 = -4\pi/3$, for all large λ; as mentioned, the modes with $X \approx X_0$ are referred to as "zero modes" [11]. Interactions between particles aggregated into a cluster lead to the formation of eigenmodes, including a family of the long-wavelength zero modes. When the cluster of particles is excited by a low-frequency applied field, so that $X(\omega) \approx X_0$, absorption is primarily due to the excitation of the zero modes and is large because of its resonance character. In contrast, for nonaggregated and well separated particles, the absorption spectrum is mostly centered in the narrow region near the center $X(\lambda_0) \approx 0$ (e.g. $\lambda_0 \approx 400$ nm for silver particles in water). In this case, there are no resonances in the long-wavelength part of the spectrum, where $X(\lambda) \approx X_0$, and, therefore, the absorption is small. Thus the formation of the long-wavelength optical modes, which accompanies particle clustering, results in enhanced far-infrared absorption.

Next we consider the localization length, $L(w_n) \equiv L_n$, characterizing a quasi-static eigenstate $|n\rangle$. The $3N$ projections of the $|n\rangle$ vector on the orthonormal basis $|i\alpha\rangle$ determine its spatial behavior. The weight with which a given nth eigenstate is localized on the ith monomer is given by $m_n(\mathbf{r}_i) \equiv m_n(i) = \sum_\alpha [\langle i\alpha|n\rangle]^2$; they are normalized by the condition $\sum_i m_n(i) = 1$. In terms of these weights, the localization length L_n of the nth eigenmode can be defined as [49,53,75]:

$$L_n \equiv L(w_n) = \sum_{i=1}^{N} m_n(i)\mathbf{r}_i^2 - \left[\sum_{i=1}^{N} m_n(i)\mathbf{r}_i\right]^2 . \tag{3.41}$$

This formula is actually a discrete function of its argument w_n. We can obtain a smooth localization function $L(X)$ by averaging $L(w_n)$ over some given interval ΔX for an ensemble of clusters

$$L(X) = \langle [K(X, \Delta X)]^{-1} \sum L(w_n) \rangle, \tag{3.42}$$

where the summation is taken over all n satisfying the condition $|X - w_n| \leq \Delta X$ and $K(X, \Delta X)$ is the number of terms in this sum. The symbol $\langle ... \rangle$ denotes here an average over an ensemble of random clusters (not to be confused with the bra- and ket-vectors).

In Fig. 3.4 the results of simulations of $L(X)$ for CCAs ($\Delta X \approx 0.6$) [53] are shown. The points indicate values of the original function $L(w_n)$ for one particular cluster while the solid line shows the result of averaging over 10 random cluster realizations.

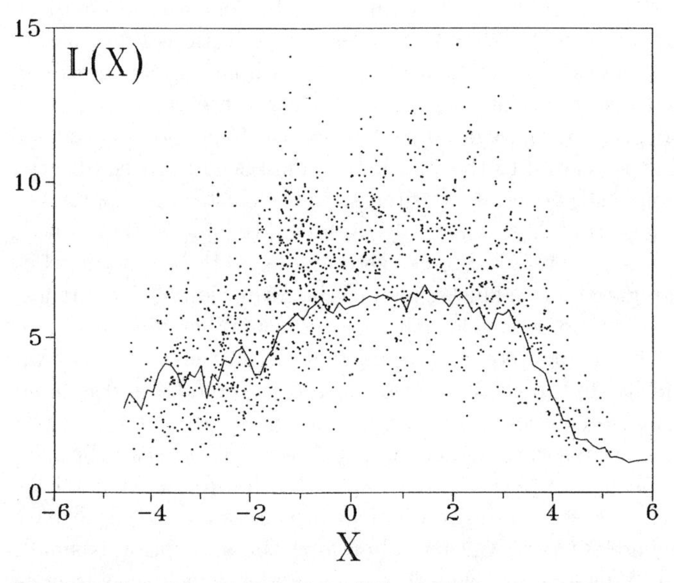

Fig. 3.4. Localization lengths $L(X)$ of dipole eigenmodes versus their eigenvalues X for CCAs. The dependence $L(X)$ averaged over an interval of $\Delta X = 0.6$ for ten random cluster realizations, containing 500 particles each, is shown by the solid line

From Fig. 3.4, we see that $L(w_n)$ exhibits large fluctuations, especially near the central point $X = 0$. There are modes that are strongly localized and some that are delocalized. The mode localization increases, on average, toward larger values of $|X|$, so that for the most localized modes $L(X)$ reduces to a dimension comparable to the size of a monomer, a.

As was emphasized in [48, 49] and mentioned above, individual eigenmodes at any given frequency are dramatically different from each other, and their sizes vary in a wide range; therefore, this localization is called inhomogeneous [48, 49]. For small values of the eigenvalues corresponding to $X \approx 0$, even chaotic behavior of the eigenmodes in fractals can be found [49]. We note that the localization length in (3.41) and (3.42) (similar to the formula used in [49]) is defined so that if a mode would consist of two or more isolated sharp peaks that are located at distances comparable to the size of a system l, then $L_n \sim l$; thus, the formula used for L_n ignores the possibility of several hot spots within a single mode.

We now compare the results of calculations with experimental data on optical extinction by fractal aggregates [53]. For a light beam propagating in a system that contains randomly distributed clusters far away from each other (so that the clusters do not interact), the intensity dependence is given by the expression $I(z) = I(0)\exp(-\sigma_e \varsigma z)$, where ς is the cluster density $\varsigma = p/[(4\pi/3)R_{\exp}^3\langle N\rangle]$ and p is the volume fraction filled by spherical particles. We introduce the extinction efficiency

$$Q_e = \frac{\langle\sigma_e\rangle}{\langle N\rangle \pi R_{\exp}^2} = \frac{4k\mathrm{Im}[\alpha]}{R_{\exp}^2}, \tag{3.43}$$

and express the intensity dependence $I(z)$ in the form

$$I(z) = I(0)\exp\left[-\frac{3}{4}Q_e p(z/R_{\exp})\right]. \tag{3.44}$$

As follows from (3.44), the extinction efficiency Q_e is the quantity that is measured in experiments on light transmission (rather than σ_e).

In [53, 74, 78] experiments were performed to measure extinction in silver colloid aggregates. In the experiments of [53] the fractal aggregates of silver colloid particles were produced from a silver sol generated by reducing silver nitrate with sodium borohydride as described in [8]. The color of fresh (non-aggregated) colloidal solution is opaque yellow; the corresponding extinction spectrum (see Fig. 3.5) peaks at 400 nm, with the halfwidth about 40 nm. Addition of some organic adsorbents (e.g. fumaric acid) promotes aggregation, and fractal colloid clusters (aggregates) are formed. When adding the fumaric acid ($0.1\,\mathrm{cm}^3$ of $0.5\,\mathrm{M}$ aqueous solution) into the colloids ($2.0\,\mathrm{cm}^3$), the colloid's color changes through dark orange and violet to dark grey over ten hours. Following the aggregation, a large wing in the long wavelength part of the spectrum appears in the extinction, as seen in Fig. 3.5. Note that in calculations shown in Fig. 3.5, the data of Fig. 3.2 was used, where X and δ are expressed in terms of λ for silver particles.

A broadening of absorption spectra of silver colloidal aggregates due to the interactions between particles constituting the aggregates is clearly seen in Fig. 3.5. The results of calculation of the absorption wing shape by the coupled dipoles method describe experimental data fairly well [53]. The calculations were performed for 500-particle CCAs (solid line with a large wing) and for 10 000-particle CCAs (circles) [53]. Clearly, the aggregation results in a large tail in the red and infrared parts of the spectrum, which is well

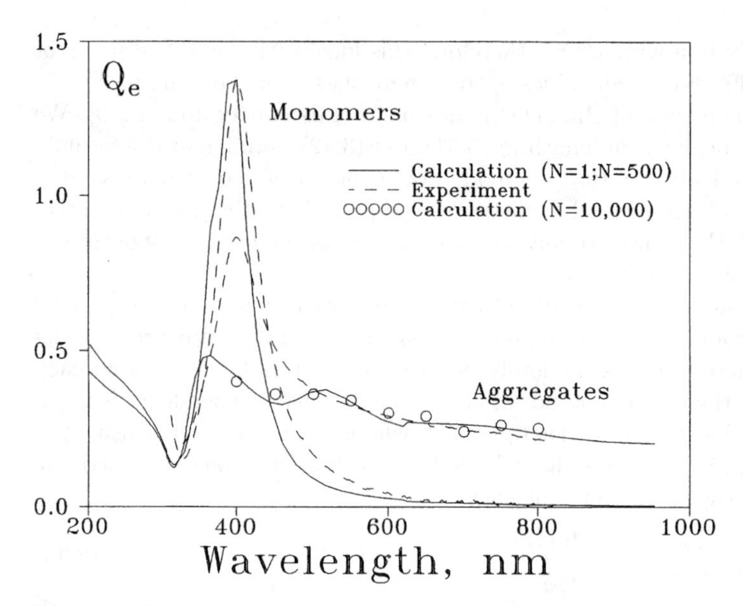

Fig. 3.5. Experimental and calculated extinction spectra of silver colloid CCAs. The theoretical spectra are presented for 500-particle and 10 000-particle CCAs

described by the simulations. This tail extends to the IR part of the spectrum and, in particular, is responsible for the enhanced IR absorption. The discrepancy in the central part of the spectrum probably occurs because in the experiments a number of particles remained nonaggregated and led to additional (not related to fractal aggregates) absorption near 400 nm.

3.4 Local-Field Enhancement in Fractals

We now discuss enhancement of local fields in small-particle composites. The parameter characterizing the enhancement of local field intensity is defined as

$$G = \langle |E_i|^2 \rangle / |E^{(0)}|^2, \tag{3.45}$$

where the angular brackets denote the averaging over an ensemble of random clusters. The enhancement G is related to linear absorption $\mathrm{Im}[\alpha(X)]$ through the exact formula that can be obtained from (3.33), (3.34), and (3.39) using the orthogonality property of the eigenmodes [46]:

$$G = \delta[1 + X^2/\delta^2]\mathrm{Im}[\alpha(X)]. \tag{3.46}$$

According to (3.46), the enhancement factor $G \approx (X^2/\delta)\mathrm{Im}[\alpha(X)]$ for $|X| \gg \delta$, i.e. it can be very large.

Note that since the fluctuations in fractals are very large, so that $\langle |E|^2 \rangle \gg \langle |E| \rangle^2$, we have $\langle |\Delta E|^2 \rangle \approx \langle |E|^2 \rangle$; therefore, in this case, G characterizes both the enhancement of local fields and their fluctuations. In other words, the larger the fluctuations, the stronger the enhancement.

Next, we consider results of numerical simulations of G for cluster-cluster aggregates (CCAs) having a fractal dimension $D \approx 1.8$, and for two nonfractal random ensembles of particles: a random gas of particles (RGP) and a close-packed sphere of particles (CPSP), as described above.

In Fig. 3.6, results of simulations for the enhancement factor G in silver CCAs in a vacuum are compared with those for nonfractal composites, namely RGP and CPSP [54]. The material optical constants for silver are taken from [89]. As can be seen in Fig. 3.6, the enhancement of local-field intensities in fractal CCAs is significantly larger than in nonfractal RGP and CPSP clusters, as was anticipated. The enhancement reaches very high values ($\sim 10^3$) at $\lambda \sim 1\mu$m and it increases further with λ. This occurs because both localization of fractal excitations and their mode quality factor ($q \sim 1/\delta \sim |\epsilon - \epsilon_h|^2/3\epsilon''\epsilon_h$) increase in the long-wavelength part of the spectrum.

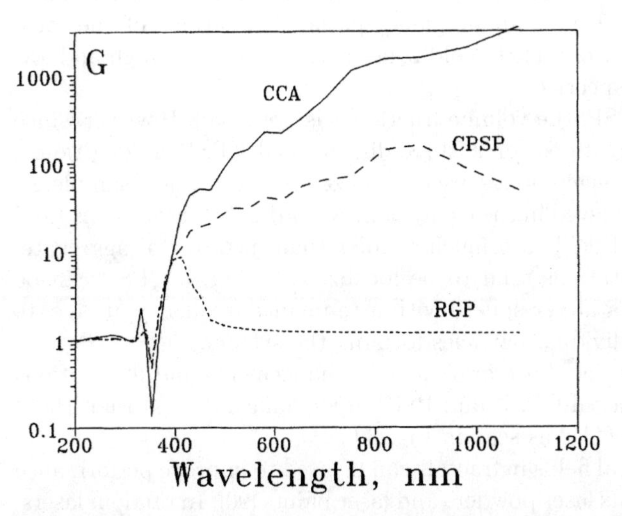

Fig. 3.6. Enhancement factors G of local field intensities plotted against wavelength λ for various random aggregates: fractal CCAs (*solid line*); a nonfractal random gas of particles (RGP) with the same volume fraction of metal as for CCAs (*short-dashed line*); and a close-packed sphere of particles (CPSP; *long-dashed line*)

We next perform a more detailed comparison between fractal small-particle composites and nonfractal inhomogeneous media. We compare CCAs first with RGP and then with CPSP. The above simulations were performed for RGP and CCA having the same volume fraction p filled by metal. The volume fraction p of particles in a fractal cluster is very small. (In fact, $p \to 0$ as $R_c \to \infty$; but p is, of course, finite for a finite number of particles.) According to the Maxwell-Garnett theory (see Sect. 1.2), there is only one resonant frequency in conventional ($d = D$) media with $p \ll 1$; the resonance is just slightly shifted from the resonance of an isolated particle occurring at $X(\omega) \approx 0$. As mentioned above, in fractal media, despite the fact that p asymptotically tends to zero, there is a high probability of finding a number of particles close to any given one – this is because the pair correlation function, $g(r) \propto r^{D-d}$, increases with decreasing distance r between particles. Thus, in fractals there is always a strong interaction of a particle with others distributed in its random neighborhood, despite $p \to 0$. As a result, there exist localized eigenmodes with distinct spatial orientations in different parts of a cluster, where the location depends on the frequency and polarization characteristics of the mode. As mentioned, some of these modes are significantly shifted to the red part of the spectrum, where their quality factors $q \sim \delta^{-1}$ are larger than that at $X(\omega) \approx 0$, for non-interacting particles. Thus, the dipole-dipole interactions of constituent particles in a fractal cluster generate a wide spectral range of resonant modes, with enhanced quality-factors and with spatial locations that are sensitive to the frequency and polarization of the applied field. The localization of optical excitations in various random parts of a cluster also brings about giant spatial fluctuations of the local fields, when one moves from hot to cold zones corresponding to high and low field-intensity areas respectively.

In the case of a CPSP, the volume fraction p is not small. However, since the dipole-dipole interactions for a three-dimensional CPSP is long-range, we can expect that the eigenmodes are delocalized over the whole sample, so that all particles are involved in the excitation. Accordingly, fluctuations (and enhancements) of local fields are much smaller than in a fractal aggregate, where the optical excitations tend to be localized. In fact, a CPSP can be roughly considered as a larger sphere with a resonance frequency at $X \approx 0$, the same as for the individual particles forming the system.

In accordance with the above arguments, enhancements and fluctuations of local fields in nonfractal CPSP and RGP are significantly less than those in the case of fractal CCAs, as seen in Fig. 3.6.

Enhancement of local fields in fractals can be used to improve performance of random lasers, such as laser powders and laser paints [90]. In random lasers, laser-like emission occurs in multiple light-scattering dielectric (or semiconductor) structures. These disordered microstructures can become alternative sources of coherent light emission. Weak scattering of light has traditionally been considered detrimental to laser action, since such scattering removes photons from the lasing mode of a conventional cavity. On the other hand, if stronger, multiple scattering occurs, photons may return to the amplification region and the amplified mode itself may consist of a multiple-scattering path [90]. By doping fractal aggregates to powder or paint lasing media, one can decrease the power of a pump needed for the lasing or in other words decrease a lasing threshold. Moreover, since the generated light can in turn also be enhanced in fractals, the intensity generated by the fractal-doped random laser is anticipated to increase.

Another interesting application is related to the possibility to control the spontaneous emission (SE) rate [91]. An ability to enhance the SE rate (the Purcell effect) would open novel avenues for physics and engineering – for instance the development of high-efficiency light emitters. Originally, Purcell considered a localized dipole in resonance with a single-cavity mode with a quality-factor Q. The SE rate in the cavity mode, referenced to the total SE rate in a homogeneous medium, is given by the Purcell figure of merit $F \sim Q\lambda^3/V$, where V is the effective medium volume. By placing a fractal aggregate in a microcavity of volume $\sim \lambda^3$, it is possible to achieve further enhancement of the zero-point fields (and thus the SE rate) because of high-quality factors of the fractal optical excitations and their localization in sub-wavelength areas. In Sect. 3.8 it will be shown that huge multiplicative enhancement can be obtained under the simultaneous, combined enhancement provided by microcavity resonators and fractal nanocomposites. Note, in particular, that a microcavity of very high resonance quality can be formed by a defect in a photonic band crystal.

3.5 Near-Field Imaging and Spectroscopy of Hot Spots

The localized optical excitations, referred to as hot spots, in fractal aggregates can be imaged using photon scanning tunneling microscopy (PSTM) providing sub-wavelength spatial resolution. For example, in [77] fractal aggregates of silver colloidal particles were prepared in solution and then deposited onto a glass prism, where the water was soaked out. The optical images were taken at different light wavelengths and polarizations. The near-field optical images shown in Fig. 3.7 exhibit spatially localized (within 150–250 nm, i.e. in sub-wavelength areas) intensity enhancement up to 10^2 for different wavelengths and polarizations. Note that a finite size of the PSTM tip results in averaging out the local enhancements over the area exceeding, in many cases, the hot spot sizes; therefore, the actual enhancement in a hot spot smaller than the tip size can be much larger than the average detected by the PSTM tip. The finite tip size also limits the spatial resolution in determining the hot spot's size.

As seen in Fig. 3.7, the spatial positions of hot spots change with both wavelength and polarization, as predicted by theory. Similar results were earlier described in [47], where first experimental observation of localization of the fractal optical excitations was reported.

In Fig. 3.8, we see the calculated spatial distribution of the local field assuming a constant tip-surface distance of 100 nm. This is somewhat larger than the tip-surface separation used experimentally. It was chosen, however, to simulate the averaging effect of the experimental tip, which is not a point tip as assumed in the calculations. The chosen 100 nm tip-surface separation

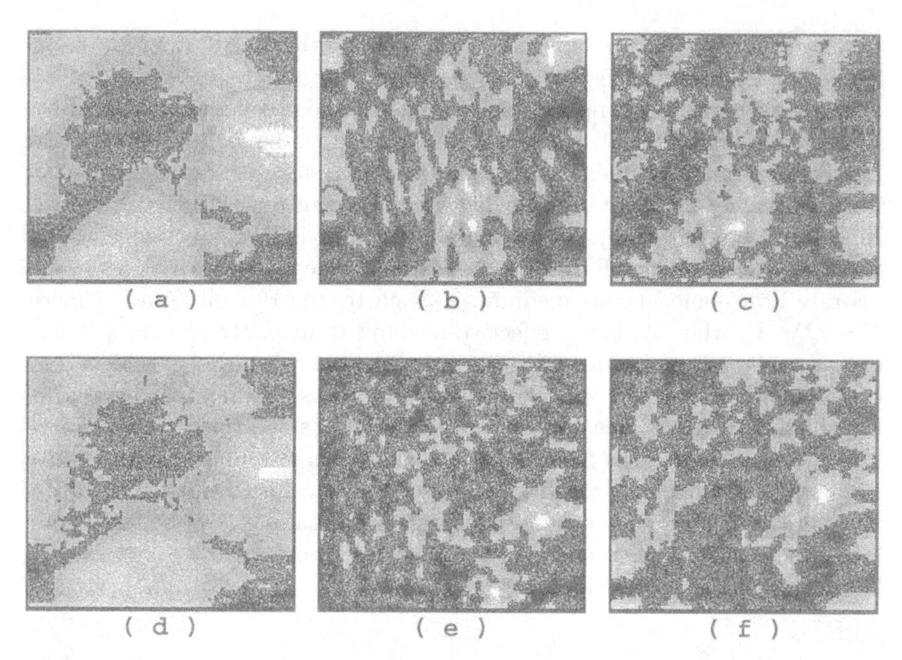

Fig. 3.7. Gray-scale topographical (**a, d**) and near-field optical (**b, c, e, f**) images ($4.4 \times 4.4\,\mu m^2$) of the same area of the sample surface taken with red s- (**b**) and p-polarized (**c**) laser beams, and with yellow s- (**e**) and p-polarized (**f**) laser beams. The angle of beam incidence was 50° for all images. The depth of the topographical images [(**a**) was taken simultaneously with (**b**), and (**d**) with (**e**)] is 85 nm. Contrast of the optical images is 98% (**b**), 88% (**c**), 99% (**e**) and 97% (**f**)

is close to the optical aperture of the tip. The pattern of calculated hot spots and their strong spatial dependence on light characteristics are very similar to that observed experimentally (see Fig. 3.7).

As mentioned, the images shown in Fig. 3.8 were purposely calculated well above the surface in order roughly to simulate the experimental situation. At that distance, the intensity contrast between the hot spots and the cold zones, although quite dramatic, is nevertheless not greater than a factor of one hundred. Very close to the surface, the local-field intensities are much more heterogeneous, so that the variation in the field intensity between the hot and cold spots can exceed a factor of 10^5. This is shown vividly in Fig. 3.9, which displays the calculated local field intensity right at the surface of a simulated silver cluster deposited onto a plane at an excitation wavelength of 1 μm. Thus, although on average the local field enhancement is in the range 10^2 to 10^3 (see Fig. 3.6), locally, in the hot spots right on the surface, the local-field enhancement can be much higher, up to a factor of 10^5 and even more in the infrared part of the spectrum.

The fact that the local enhancement in the hot spots can exceed the average enhancement by many orders of magnitude is extremely important

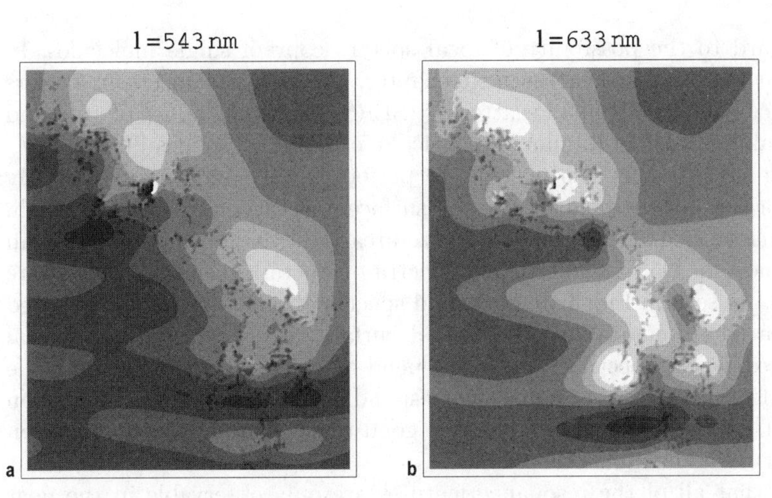

Fig. 3.8. Computer-generated optical image of a cluster taken in the constant-distance mode for s-polarized light with $\lambda = 543$ nm (**a**) and 633 nm (**b**) at 50° angle of laser beam incidence. The size of the sample is approximately $1.2 \times 0.7 \mu m^2$. The topography of the cluster can be seen as a set of small black points

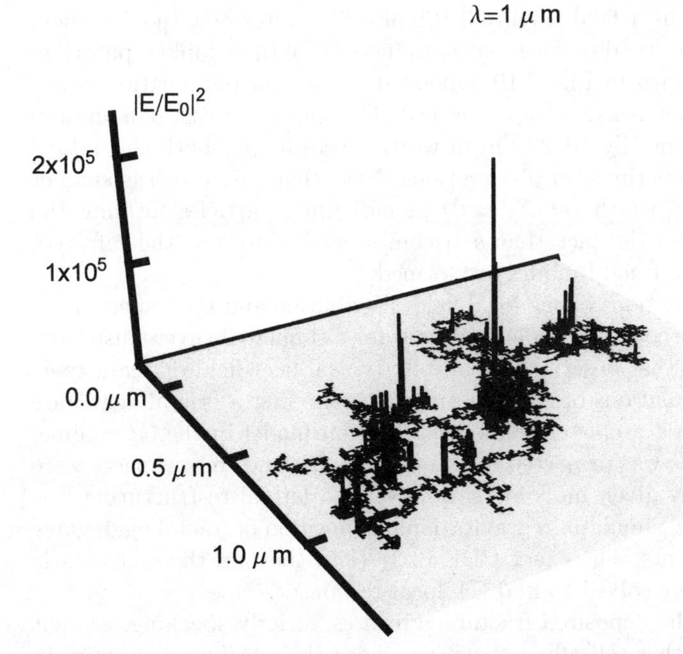

Fig. 3.9. Calculated field distributions right on the surface of silver fractal aggregates deposited on a plane

with regard to the possibility of local spectroscopy of single molecules. In particular, as discussed in the next chapter, the local enhancement factors for surface-enhanced Raman scattering (SERS) can reach values of 10^{12} and more, making possible Raman spectroscopy of single molecules [92–94].

Using PSTM, it is also possible to perform local nanospectroscopy. By parking a tip at some point above the surface and varying light wavelength, the resonance frequencies of the nm-size area right undernearth the tip can be obtained. In [95], first near-field spectra of fractal silver aggregates were reported. As seen in Fig. 3.10, near-field spectra observed (a) and calculated (b) at various points above the fractal surface consist of several spectral resonances, whose spectral locations depend on the sample site probed. The more distinct of these resonances correspond to individual surface plasmon modes of a fractal aggregate. There is good qualitative agreement between the observed and simulated spectra.

Note that all of the resonance features are only observable in the near field. In the far field, images and spectra are observed in which hot spots and the spectral resonances are averaged out. This effect is illustrated in the bottom picture in Fig. 3.10b: the average far-zone spectra for absorption (Q_a), extinction (Q_e) and scattering (Q_s) do not show resonance structures that can be seen in the near field. Figure 3.10b also illustrates how the near-field resonance structures tend to wash out with increase of tip-sample separation.

The spectra shown in Fig. 3.10 depend markedly on the location above the film at which the near-field tip is parked. The spectra consist of a number of narrow bands typically 10–20 nm in width, according to both theoretical calculations and experimental observations. Note that this width is smaller than the resonance width (at $X \approx 0$) of individual particles forming the fractal. This reflects the fact that δ becomes smaller toward the infrared, where the collective fractal modes are formed.

The above simulations of the local field distribution and their spectra are done by solving the coupled-dipole equations for a simulated silver cluster deposited on a plane that describes accurately typical near-field optical experiments [95]. In simulations of the deposited clusters, first a fractal aggregate was generated, using a cluster-cluster aggregation model in the $3d$ volume. Then, the aggregate was projected onto a plane so that no empty spaces were left underneath any given monomers; however, no lateral restructuring was allowed. This model simulates a gravitational deposition of fractal aggregates in PSTM experiments. The exact CDE (3.2)–(3.5) (beyond the quasi-static approximation) were solved to find the local responses.

We note that the deposited fractal aggregates, strictly speaking, are not self-similar, but rather self-affine structures since their scaling properties in the (x, y) plane and in the normal z direction are, of course, different. The self-affine films will be considered in detail in the next chapter.

The enhanced local fields result in enhancements of the optical processes considered below. Based on the simulations presented above, one anticipates

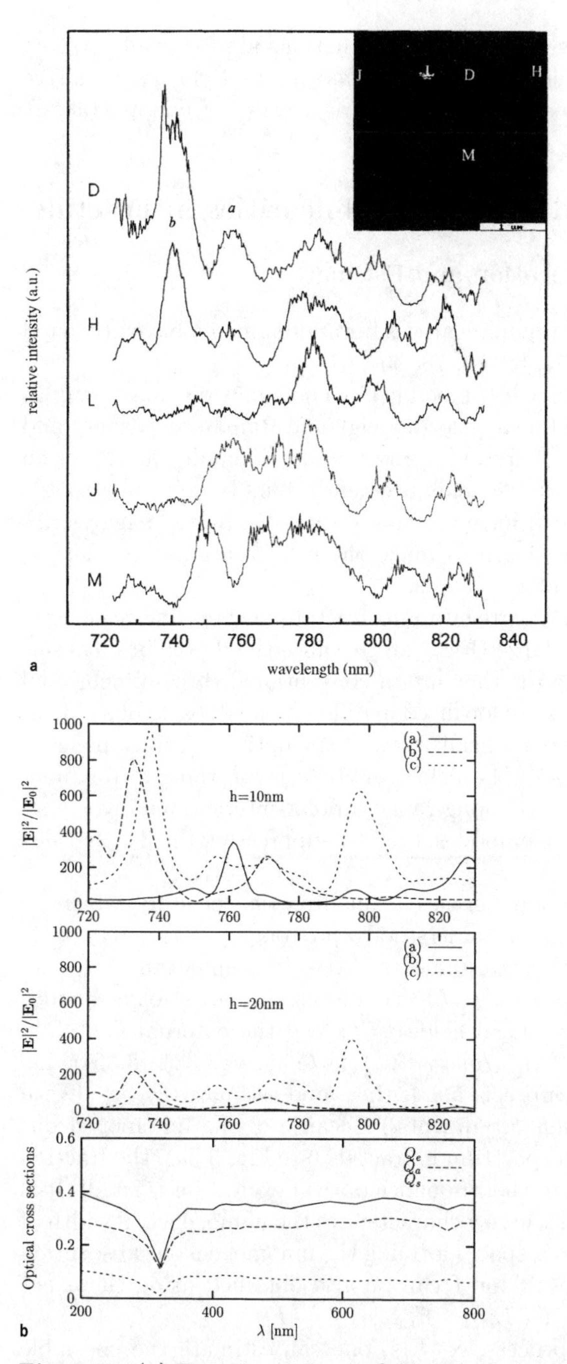

Fig. 3.10. (**a**) Experimental near-field optical spectra at five different spots (shown in the inset as J, L, D, H, and M) on silver fractal aggregates deposited on a plane. Spectra were taken with a Ti/sapphire laser. (**b**) Calculated near-field optical spectra at three different spots (a, b, and c) for two different tip-sample separations (10 nm and 20 nm). The bottom picture within (**b**) shows the average spectra for extinction, absorption and scattering. ($E_0 \equiv E^{(0)}$ is the applied field)

that in fractal composites, where the field fluctuations are especially strong, enhancements of optical nonlinearities can be very large. Below, we analyze various enhanced optical phenomena in fractal aggregates of nanoparticles.

3.6 Surface-Enhanced Optical Nonlinearities in Fractals

3.6.1 Qualitative Consideration and Estimates

Large enhancements of optical nonlinearities in random media have attracted much attention [11, 20, 45, 54, 69–74, 78, 96–101].

The clustering of small particles embedded in a host material may result in a giant enhancement of both linear (e.g. Rayleigh and Raman scattering) and nonlinear (four-wave mixing, harmonic generation, and nonlinear reflection and absorption) optical effects. The enhancement occurs because of strongly fluctuating local fields that can have very large values in particle aggregates (see Figs. 3.6 and 3.9). Nonlinearities make these fluctuations even larger, leading eventually to giant enhancements.

As mentioned, if particles aggregate into fractal clusters, fluctuations of the local fields are especially large (Fig. 3.6), because the dipole interactions in fractals are not long-range (as they are in conventional three-dimensional media) and optical excitations are localized in different small parts of a cluster with various random structures. Localization of the optical excitations leads to strong spatial fluctuations of the fields. In contrast, in random but non-fractal ensembles of particles, the long-range dipolar interactions involve all particles in the excitation of eigenmodes, thereby suppressing the fluctuations (see Fig. 3.6).

Enhancement of optical nonlinearities in small-particle clusters, which we consider in detail below, can be understood and roughly estimated using the following simple and robust arguments. Consider enhancement for an arbitrary nonlinear optical process $\propto E^n$. As discussed above, for resonance dipole eigenmodes in fractals, local fields E_i exceed the external field $E^{(0)}$ by the factor $E_{i,\mathrm{res}}/E^{(0)} \sim |\alpha_0^{-1}/\delta| = |X + \mathrm{i}\delta|/\delta$ (see (3.33)–(3.36)), i.e. $\sim |X|/\delta$ for $|X| \gg \delta$. (In Subsect. 5.3.4, the same estimate $E_{\mathrm{res}} \sim |X|/\delta$ will be obtained from different arguments.) Because of the inhomogeneous broadening of the absorption spectrum in fractals (see Fig. 3.3a), the fraction f of the monomers involved in the resonance optical excitation is small. It is estimated as $f \sim \delta \mathrm{Im}[\alpha(X)]$, where δ characterizes the homogeneous width of individual resonances (in the X space) forming the inhomogeneous absorption spectrum $\mathrm{Im}[\alpha(X)]$. This result for f can be also obtained using the exact formula (3.46). Note also that $\langle |E_i| \rangle \sim E_{i,\mathrm{res}} \times f \sim |E^{(0)}|$.

For a nonlinear optical process, $\propto |E|^n$, one can estimate the ensemble average of the enhancement, $\langle |E_i/E^{(0)}|^n \rangle$, as the resonance value $|E_i/E^{(0)}|^n_{\mathrm{res}}$ multiplied by the fraction of the resonant modes f, i.e. by the fraction of particles involved in the resonance excitation. (The sign $\langle ... \rangle$ stands here for the ensemble averaging.) Thus we arrive at the following estimate for nonlinear enhancement:

$$\langle |E_i/E^{(0)}|^n \rangle \sim |X|^n \delta^{-n} \times \delta \mathrm{Im}[\alpha(X)] \sim |X|^n \delta^{1-n} \mathrm{Im}[\alpha(X)], \qquad (3.47)$$

which is $\gg 1$ for $n > 1$. Since this is only a rough estimation, an adjustable constant C should, in general, be added as a pre-factor. Formula (3.47) for $n = 2$ reproduces the exact result given in (3.46).

The nonlinear dipole amplitude can be enhanced along with the linear local fields, provided that generated in a nonlinear process frequency lies within the spectral region of the fractal eigenmodes. In this case, for enhancement of incoherent processes, such as Raman scattering (G_{RS}) and optical Kerr effect (G_{K}), $n = 4$ and we obtain from (3.47):

$$G_{\mathrm{RS,K}} \sim C_{\mathrm{RS,K}} X^4 \delta^{-3} \mathrm{Im}[\alpha(X)], \qquad (3.48)$$

where constant pre-factors C_{RS} and C_{K} can be different for G_{RS} and G_{K}.

For coherent processes, the resultant enhancement $\sim |\langle |E_i/E^{(0)}|^n \rangle|^2$. Also, for quasi-degenerate four-wave mixing (FWM), additional enhancement of the generated nonlinear amplitudes oscillating at almost the same frequency as the applied field should be included. Thus (recall that, according to (2.2), $G_{\mathrm{FWM}} = |G_{\mathrm{K}}|^2$) the enhancement factor is given by

$$G_{\mathrm{FWM}} \sim C_{\mathrm{FWM}} X^8 \delta^{-6} \left(\mathrm{Im}[\alpha(X)]\right)^2 . \qquad (3.49)$$

All these estimates are in agreement with results of the analytical and numerical calculations discussed in the following sections of this chapter.

Note that other optical phenomena, such as hyper-Raman scattering, two-photon absorption, harmonic generation, etc., can also be enhanced in small-particle composites. For example, fluorescence (from molecules adsorbed on a small-particle aggregate) following the two-photon absorption by the aggregate is enhanced by the factor: $G_{\mathrm{F}} \sim \langle |E_i/E^{(0)}|^4 \rangle \sim |\alpha_0|^{-4} \langle |\alpha_i|^4 \rangle \propto \delta^{-3}$. Using the above simple considerations, one can roughly estimate enhancement for an arbitrary optical process.

As shown in early papers [45], enhancement is typically stronger for multi-photon processes that include annihilation of at least one of the photons of incident waves in an elementary act of photon scattering. For example, the Kerr optical effect and four-wave mixing are enhanced stronger than processes of high-harmonic generation. The photon annihilation for one of the applied fields means that the corresponding frequency enters the nonlinear susceptibility with a negative sign, which formally means photon subtraction (annihilation) in the photon balance. A lower enhancement for processes, where the incident wave photons are added in one nonlinear scattering act to form the photon of a generated wave, can be explained by the destructive interference between the fields generated from different parts of a system. Formally, this is a consequence of the difference in averaging different processes over the mode distribution (see also Chap. 5). For the processes without photon subtraction, all the poles, representing the system resonances, are located in the same complex semiplane, whereas the processes including at least one photon subtraction always have poles in different complex semiplanes. In the latter case, the averaging over the mode distributions leads to a larger total enhancement.

We also note that the above qualitative estimates rely significantly on the fact that optical excitations are localized, which is the case for fractals. In these estimates we have assumed that different modes of a particle aggregate are spatially located in different small parts of the aggregate and resonate independently, resulting in particularly strong inhomogeneous broadening – compare the spectra of fractal and nonfractal aggregates in Figs. 3.3a and 3.3b respectively. The explicit dependence on the fractal morphology enters in the above formulas through the absorption $\mathrm{Im}[\alpha(X)]$, which depends, of course, on the fractal dimension D and experiences dramatic changes at the transition from nonfractal, $D = d = 3$, to fractal, $D < d$, media.

Thus, the fractal morphology of small-particle aggregates results in very strong local fields associated with localized optical excitations in fractals; the large and strongly fluctuating local fields lead to enhancement of optical phenomena in fractal aggregates. Below, we consider a rigorous theory of surface-enhanced optical processes in fractals.

3.6.2 Enhanced Raman and Rayleigh Scattering

We first consider in this subsection enhancement of Raman scattering G_{RS} resulting from aggregation of particles into fractal clusters. We assume that each monomer of a cluster, apart from linear polarizability α_0, also possesses Raman polarizability ζ. This means that the exciting field $\mathbf{E}^{(0)}$, applied to an isolated monomer, induces dipole moment $\mathbf{d}^{\mathrm{s}}$ oscillating with the Stokes-shifted frequency ω_{s}. To avoid unnecessary complications, we suppose ζ to be a scalar; this gives $\mathbf{d}^{\mathrm{s}} = \zeta\mathbf{E}$. The Raman polarizability may either be due to the polarizability of a monomer itself or to an adsorbant Raman-active molecule bound to the monomer.

Note that in the red and infrared part of the spectrum the enhancement of the local fields in nonaggregated metal particles is negligible in comparison with the enhancement in aggregates (clusters) of particles. Therefore, we can define enhancement of Raman scattering (and other optical processes considered below) as the one that results from aggregation of particles into clusters.

We consider spontaneous Raman scattering (RS), which is an incoherent optical process. This means that Raman polarizabilities ζ_i corresponding to different monomers include uncorrelated random phases:

$$\langle \zeta_i^* \zeta_j \rangle = |\zeta|^2 \delta_{ij}. \tag{3.50}$$

This feature constitutes the principal distinction between ζ and linear polarizability α. It ensures that there exists no interference of the Stokes waves generated by different monomers.

As was pointed out above, when monomers are the constituents of a cluster, the field acting upon the ith monomer is the local field $\mathbf{E}_i$ rather than the external field $\mathbf{E}^{(0)}$. Also, dipole interactions of the monomers at the Stokes-shifted frequency ω_{s} should be included. These interactions occur through the linear polarizability $\alpha(\omega_{\mathrm{s}}) = \alpha^{\mathrm{s}}$ at the Stokes frequency ω_{s}. Taking these arguments into account, we can write the following system of equations

$$d_{i\alpha}^{\mathrm{s}} = \zeta_i E_{i\alpha} + \alpha_0^{\mathrm{s}} \sum_{j\beta} (i\alpha|W|j\beta) d_{j\beta}^{\mathrm{s}}, \tag{3.51}$$

where α_0^{s} is the linear polarizability of an isolated monomer at the Stokes-shifted frequency ω_{s}.

Note that, to avoid confusion with the sign for the statistical averaging, we use hereafter for eigenstates the notations $|...)$ and $(...|$ instead of $|...\rangle$ and $\langle...|$ used above.

The total Stokes dipole moment $\mathbf{D}^{\mathrm{s}}$ found by solving (3.51) is [52]:

$$D_\alpha^{\mathrm{s}} = \sum_i d_{i\alpha}^{\mathrm{s}} = Z_{\mathrm{s}} Z \sum_j \zeta_j \alpha_{j,\beta\alpha}^{\mathrm{s}} \alpha_{j,\beta\beta'} E_{\beta'}^{(0)}, \tag{3.52}$$

where $Z_{\mathrm{s}} = (\alpha_0^{\mathrm{s}})^{-1}$, $\alpha_i^{\mathrm{s}} \equiv \alpha_i(X_{\mathrm{s}})$, and α_i are the local polarizabilities given by (3.34); these local polarizabilities are simply proportional to the local fields E_i since $d_{i\alpha} = \alpha_{i,\alpha\beta} E_\beta^{(0)} \equiv \alpha_0 E_{i\alpha}$, where $E^{(0)}$ is the applied external field. The RS enhancement is defined as [52]:

$$G_{\mathrm{RS}} = \frac{\langle |\mathbf{D}_{\mathrm{s}}|^2 \rangle}{N|\zeta|^2 |\mathbf{E}^{(0)}|^2}. \tag{3.53}$$

The formulas (3.51)–(3.53) are exact and valid for any cluster of particles. If the Stokes shift is so large that the Raman-scattered light is well out of the absorption band of the cluster, the polarizability α_i^{s} in (3.52) and (3.53) can be approximated as $\alpha_{i,\alpha\beta}^{\mathrm{s}} \approx Z_{\mathrm{s}}^{-1} \delta_{\alpha\beta}$, and then enhancement (3.53) acquires the following form after averaging over orientations [52]:

$$\begin{aligned} G_{\mathrm{RS}} &= |Z|^2 \frac{1}{3N} \left\langle \sum_i |\alpha_{i,\alpha\beta}|^2 \right\rangle = \left\langle \left| \frac{E_i}{E^{(0)}} \right|^2 \right\rangle \\ &= \delta \left(1 + X^2/\delta^2\right) \mathrm{Im}[\alpha(X)], \end{aligned} \tag{3.54}$$

where we used (3.33)–(3.36) and (3.46). Thus, if the Raman-scattered light does not interact with the cluster, the enhancement is simply proportional to the mean of the local fields squared [52].

However, the more interesting case occurs when the Stokes shift is small, so that the Stokes amplitudes are also enhanced. Then, the general expression (3.53) is needed. After averaging over orientations, we obtain

$$G_{\mathrm{RS}} = \frac{(X^2 + \delta^2)^2}{3} \langle \mathrm{Tr}[(\hat{\alpha}_i^{\mathrm{T}} \hat{\alpha}_i \hat{\alpha}_i^{\mathrm{T}*} \hat{\alpha}_i^*] \rangle. \tag{3.55}$$

where the T symbol in $\hat{\alpha}^{\mathrm{T}}$ denotes a transposition of the matrix $\hat{\alpha}$, so that, for example, $(\hat{\alpha}_j^{\mathrm{T}}\hat{\alpha}_j)_{\alpha\beta} \equiv \alpha_{j,\alpha'\alpha}\alpha_{j,\alpha'\beta}$.

For the non-resonant case $|X| \gg |w_n|$, we have $\alpha_i = \alpha_0$ and, therefore, $G_{\mathrm{RS}} = 1$ in (3.55).

According to (3.55), enhancement of Raman scattering is determined by the averaging of the enhanced local field to the fourth power (cf. (2.3)), i.e.

$$G_{\mathrm{RS}} \sim \left\langle \left| \frac{E_i}{E^{(0)}} \right|^4 \right\rangle \sim |\alpha_0^{-1}|^4 \langle |\alpha_i|^4 \rangle. \tag{3.56}$$

For further consideration we represent the RS enhancement in the following form

$$G_{\mathrm{RS}} = \frac{(X^2+\delta^2)^2}{3} \langle g_{\mathrm{RS}}(X) \rangle, \tag{3.57}$$

with $g_{\mathrm{RS}}(X)$ given by

$$g_{\mathrm{RS}} \equiv \mathrm{Tr}[(\hat{\alpha}_i^{\mathrm{T}}\hat{\alpha}_i\hat{\alpha}_i^{\mathrm{T}*}\hat{\alpha}_i^*] = \sum_{nmlk} K_{nmlk}\Lambda_n\Lambda_m\Lambda_l^*\Lambda_k^*, \tag{3.58}$$

where $\Lambda_n \equiv [(w_n - X) - i\delta]^{-1}$ and K_{nmlk} is defined as

$$\begin{aligned} K_{nmlk} \equiv \sum_{jj'j''j'''} & (i\alpha|n)(n|j\beta)(i\alpha|m)(m|j'\beta') \\ & \times (i\beta''|l)(l|j''\beta')(i\beta''|k)(k|j'''\beta). \end{aligned} \tag{3.59}$$

For further analysis we also introduce R_{nm} defined as

$$R_{nm} \equiv \frac{1}{2i\delta + (w_n - w_m)}. \tag{3.60}$$

The factor K_{nmlk} satisfies the sum rule: $\sum_{n,m,l,k} K_{nmlk} = 3$.

The function g_{RS} in turn satisfies the following relation

$$\int_{-\infty}^{\infty} g_{\mathrm{RS}}(X)\mathrm{d}X = 4\pi \mathrm{Im} \sum_{nmlk} K_{nmlk} R_{nk} R_{ml} R_{nl}. \tag{3.61}$$

Furthermore, the product of the Λ factors in (3.58) can be rewritten as

$$\begin{aligned} \Lambda_n\Lambda_m\Lambda_l^*\Lambda_k^* = R_{ml}R_{nk}\{ & (\Lambda_n\Lambda_m + \Lambda_l^*\Lambda_k^*) \\ & - R_{nl}(\Lambda_n - \Lambda_l^*) - R_{mk}(\Lambda_m - \Lambda_k^*)\}. \end{aligned} \tag{3.62}$$

We first find $g_{\mathrm{RS}}^{\mathrm{r}}(X)$ due to the "resonance" difference of eigenmodes, i.e. we find the contribution due to only such modes that $|w_n - w_m| \ll \delta$ in (3.58). In this case, for the ensemble average, we obtain

$$\langle g_{\mathrm{RS}}^{\mathrm{r}}(X) \rangle = \frac{3}{2\delta^3} \mathrm{Im}[\alpha(X)]. \tag{3.63}$$

The sum rule for the ensemble-average resonant contribution has the form

$$\int_{-\infty}^{\infty} \langle g_{\mathrm{RS}}^{\mathrm{r}}(X) \rangle \mathrm{d}X = \frac{3\pi}{2\delta^3}. \tag{3.64}$$

We conjecture that the spectral dependence predicted by the "resonance" approach is correct in the general case so that

$$G_{\mathrm{RS}} = C_{\mathrm{RS}} \frac{(X^2+\delta^2)^2}{\delta^3} \mathrm{Im}[\alpha(X)], \tag{3.65}$$

where we allow C_{RS} to be an adjustable parameter. Note that this formula is in agreement with the above estimate (3.48).

In Fig. 3.11 the results of simulations of G_{RS} defined in (3.55) are shown for negative X (results for positive X are similar). The solid line in Fig. 3.11 gives the enhancement found from (3.65), with C_{RS} obtained from the relation: $G_{\mathrm{RS}}\delta^3 \approx 1$ in the maxima occurring at $X \approx \pm 4$. The dashed line represents a power-law fit for results of the simulations for $G_{\mathrm{RS}}\delta^3$ with $\delta = 0.05$ in the interval $0.1 \le |X| \le 3$. The exponent obtained (4.07 ± 0.70) is close to value 4. Note that for $|X| < 4$ the dependence associated with a pre-factor (in this case, X^4) dominates the relatively weak (in this interval) spectral dependence of $\mathrm{Im}\alpha(X)$.

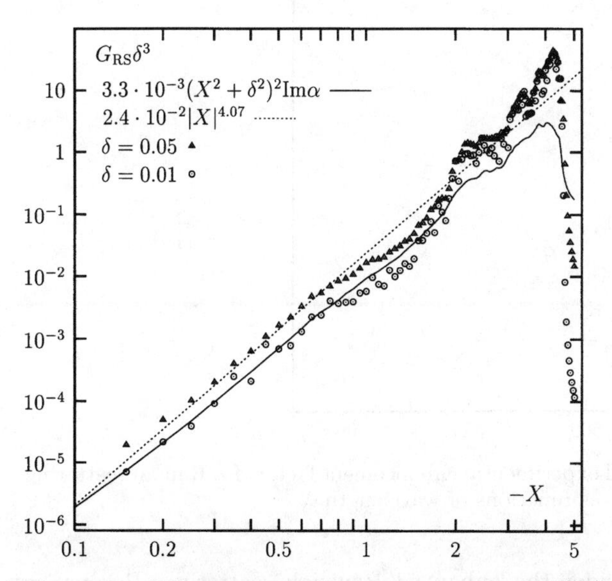

Fig. 3.11. Enhancement of Raman scattering in CCAs G_{RS} (multiplied by loss-parameter cubed δ^3) for negative X, corresponding to the visible and infrared parts of the spectrum (see Fig. 3.2a)

As seen in Fig. 3.11, the product $G_{\mathrm{RS}}\delta^3$ does not depend systematically on δ in the important region close to the maximum, and its value there is of an order of one. Thus, we conclude that the strong enhancement of Raman scattering $G_{\mathrm{RS}} \sim \delta^{-3}$, resulting from aggregation of particles into fractal clusters, can be obtained.

In Fig. 3.12, surface-enhanced RS data obtained for silver colloid solution in experiments of [102] is compared with the G_{RS} calculated with the use of (3.65). (The values of X and δ at different λ were found using the data of [89]; see Fig. 3.2.) Note that only the spectral dependence of G_{RS} is informative in this figure since only relative values of G_{RS} were measured in [102]. The experimental data presented in Fig. 3.12 is normalized by setting $G_{\mathrm{RS}} = 28\,250$ at 570 nm, which is a reasonable value. Clearly, the present theory successfully explains the huge enhancement accompanying aggregation of particles into fractals and the observed increase of G_{RS} towards the red part of the spectrum. The strong enhancement towards the red occurs because the local fields associated with collective dipolar modes in CCAs become significantly larger in this part of the spectrum (see Fig. 3.6).

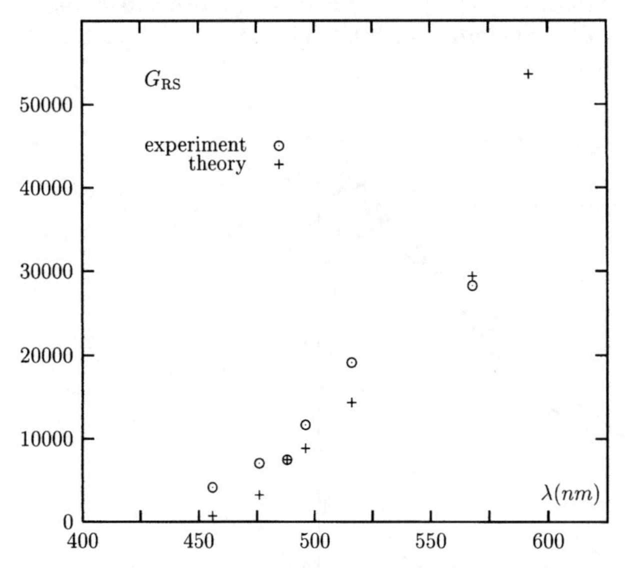

Fig. 3.12. Theoretical and experimental enhancement factors for Raman scattering in silver colloid aggregates as functions of wavelength λ

Now we briefly consider the enhanced Rayleigh scattering. Resonance Rayleigh scattering by fractal clusters was studied by Shalaev et al. [66]. The authors showed that the scattering cross-section of small-particle aggregates has the form [66]

$$\sigma_s = \frac{2\pi}{15} k^4 N (K_1 \langle \mathrm{Tr}(\hat{\alpha}^i \hat{\alpha}^{i'*}) \rangle + K_2 \langle \mathrm{Tr}(\hat{\alpha}^i) \mathrm{Tr}(\hat{\alpha}^{i'*}) \rangle), \tag{3.66}$$

$$K_1 = \begin{cases} C(kR_0)^{-D}(\frac{7}{2-D} - \frac{2}{4-D} + \frac{2}{6-D}), & \text{if } (D<2); \\ \frac{7}{2}(kR_0)^{-2} \ln((kR_0)^2 N), & \text{if } (D=2); \\ \frac{7}{2}\frac{D}{D-2}(kR_0)^{-2} N^{1-D/2}, & \text{if } (D>2). \end{cases} \tag{3.67}$$

$$K_2 = \begin{cases} C(kR_0)^{-D}(\frac{1}{2-D} - \frac{6}{4-D} + \frac{6}{6-D}), & \text{if } (D < 2); \\ \frac{1}{2}(kR_0)^{-2}\ln((kR_0)^2 N), & \text{if } (D = 2); \\ \frac{1}{2}\frac{D}{D-2}(kR_0)^{-2}N^{1-2/D}, & \text{if } (D > 2), \end{cases} \quad (3.68)$$

Here $C = D\Gamma(D-1)2^{1-D}\cos\frac{\pi}{2}(D-2)$ and $\Gamma(...)$ is the Gamma function.

The scattering enhancement factor F_{R} is defined as $F_{\mathrm{R}} = \sigma_{\mathrm{s}}/N\sigma_{\mathrm{s}}^{(0)}$, where $\sigma_{\mathrm{s}}^{(0)}$ is the single-particle scattering cross-section, $\sigma_{\mathrm{s}}^{(0)} = \frac{8\pi}{3}k^4|\alpha_0|^2$. From (3.66) we have

$$F_{\mathrm{R}} = \frac{1}{20}|\alpha_0|^{-2}(K_1\langle\mathrm{Tr}(\hat{\alpha}^i\hat{\alpha}^{i'*})\rangle + K_2\langle\mathrm{Tr}(\hat{\alpha}^i)\mathrm{Tr}(\hat{\alpha}^{i'*})\rangle). \quad (3.69)$$

In the limit of non-resonant scattering, (3.69) reduces to the result obtained first by Berry and Percival [57], namely $F_{\mathrm{R}} = \frac{3}{20}(K_1 + 3K_2)$. Note that the resonance Rayleigh scattering is proportional to the second-order field moment at two different points $\langle E(\mathbf{r})E(\mathbf{r}')\rangle$. The correlator is $\propto \delta^{-1}$ and can be very large in metal fractal aggregates [66]. In particular, this might be important for photonic crystals and may also result in favorable conditions for observation of Anderson light localization [103, 104].

3.6.3 Nearly Degenerate Four-Wave Mixing

Four-wave mixing (FWM) is a four-photon process determined by following nonlinear polarizability [55]. Thus

$$\beta^{(3)}_{\alpha\beta\gamma\delta}(-\omega_{\mathrm{s}}; \omega_1, \omega_1, -\omega_2), \quad (3.70)$$

where $\omega_{\mathrm{s}} = 2\omega_1 - \omega_2$ is the generated frequency, and ω_1 and ω_2 are the frequencies of the applied waves. As discussed in the previous chapter, coherent anti-Stokes Raman scattering (CARS) is an example of FWM, where two ω_1 photons are converted into ω_2 and ω_{s} photons. In the degenerate FWM (DFWM), which is used for optical phase conjugation (OPC), all waves have the same frequency ($\omega_{\mathrm{s}} = \omega_1 = \omega_2$). Below, we consider the DFWM process, where the total applied field is $\mathbf{E}^{(0)} = \mathbf{E}^{(1)} + \mathbf{E}'^{(1)} + \mathbf{E}^{(2)}$, with $\mathbf{E}^{(1)}$ and $\mathbf{E}'^{(1)}$ being the amplitudes of the two (typically, oppositely directed) pump beams and $\mathbf{E}^{(2)}$ being the amplitude of the probe beam.

The polarizability $\beta^{(3)}$ that results in DFWM leads also to nonlinear refraction and absorption (to be considered below) and is associated, in general, with the Kerr optical nonlinearity. For coherent effects, including the ones discussed in this section, the averaging should be performed for a generated field amplitude, i.e. with nonlinear polarization. Note also that the nonlinear polarizability $\beta^{(3)}$ can be associated either with monomers forming a cluster or with molecules adsorbed on the monomers.

We will now consider enhancement of DFWM that accompanies aggregation of particles into clusters. Let particles first be randomly embedded in a linear host medium so that there is no clustering. We assume that the volume fraction p that the particles occupy is small and interparticle interactions can be neglected. Then let particles aggregate in many random clusters that are relatively far from each other (i.e. the intercluster interactions are still negligible). Thus, after the aggregation we can obtain a mixture of many clusters (each cluster may consist of thousands of particles). The average volume fraction p filled by particles clearly remains the same; however, the particles within one cluster now strongly interact via light-induced dipolar fields. This scenario of aggregation occurs, for example, in a silver colloid solution. In that case, it is first necessary to produce a silver sol (non-aggregated particles in solution), for example, by reducing silver nitrate with sodium borohydride [8]. Addition of an adsorbent (like phtalizine) promotes aggregation, resulting in fractal colloid clusters with fractal dimension $D \approx 1.78$.

To be specific, we assume below that particles themselves possess $\beta^{(3)}$ and the enhancement results from their clustering. However, all main conclusions remain the same for the case when a nonlinear signal originates from molecules adsorbed on the surface of metal aggregates.

The orientation-averaged nonlinear polarizability in an isotropic medium can be expressed, in general, through two independent scalar functions f_{s} and f_{a} as [55]

$$\langle \beta^{(3)}_{\alpha\beta\gamma\delta} \rangle_0 = f_{\mathrm{s}} \Delta^{+}_{\alpha\beta\gamma\delta} + f_{\mathrm{a}} \Delta^{-}_{\alpha\beta\gamma\delta} \tag{3.71}$$

$$\Delta^{+}_{\alpha\beta\gamma\delta} = \frac{1}{3}\{\delta_{\alpha\beta}\delta_{\gamma\delta} + \delta_{\alpha\gamma}\delta_{\beta\delta} + \delta_{\alpha\delta}\delta_{\beta\gamma}\}, \tag{3.72}$$

$$\Delta^{-}_{\alpha\beta\gamma\delta} = \frac{1}{3}\{\delta_{\alpha\beta}\delta_{\gamma\delta} + \delta_{\alpha\gamma}\delta_{\beta\delta} - 2\delta_{\alpha\delta}\delta_{\beta\gamma}\}, \tag{3.73}$$

where the sign $\langle ... \rangle_0$, denotes the averaging over orientations. The terms $f_{\mathrm{s}}\Delta^{+}$ and $f_{\mathrm{a}}\Delta^{-}$ are totally and partially symmetric parts of $\beta^{(3)}_{\alpha\beta\gamma\delta}$ respectively.

When a cluster consists of monomers, the field acting upon them is the local field $\mathbf{E}_i$ rather than the applied field $\mathbf{E}^{(0)}$. Also, the dipolar interaction of *nonlinear* dipoles at the generated frequency should be taken into consideration. Taking these arguments into account, we can write the following system of equations for the light-induced nonlinear dipoles

$$d^{\mathrm{NL}}_{i,\alpha} = 3\beta^{(3)}_{\alpha\beta\gamma\delta} E_{i,\beta} E_{i,\gamma} E^{*}_{i,\delta} + \alpha(\omega_{\mathrm{s}}) \sum_j W_{ij,\alpha\beta} d^{\mathrm{NL}}_{j,\beta}, \tag{3.74}$$

where the pre-factor 3 represents the degeneracy factor that gives the number of distinct permutations of frequencies ω, ω, and $-\omega$ [55]. Note that (3.74) is an analog of the coupled-dipole equation (3.25) for nonlinear dipoles.

The nonlinear polarization $P^{(3c)}$ of an isotropic (on average) composite material consisting of particles aggregated into clusters has the form [55]:

$$\mathbf{P}^{(3c)}(\omega) = A\mathbf{E}^{(0)}(\mathbf{E}^{(0)} \cdot \mathbf{E}^{(0)*}) + \frac{1}{2} B\mathbf{E}^{(0)*}(\mathbf{E}^{(0)} \cdot \mathbf{E}^{(0)}), \tag{3.75}$$

where A and B are given by

$$A = \frac{2}{3}(F_{\mathrm{s}} + F_{\mathrm{a}})pv_0^{-1}, \qquad B = \frac{2}{3}(F_{\mathrm{s}} - 2F_{\mathrm{a}})pv_0^{-1}, \tag{3.76}$$

v_0 is the volume of a particle ($v_0 = (4\pi/3)R_{\mathrm{m}}^3$, for a sphere), and F_{s} and F_{a} are totally and partially symmetric parts of the average cluster polarizability $\langle\beta^{(3c)}_{\alpha\beta\gamma\delta}\rangle$ having the form similar to (3.71)–(3.73), so that

$$\langle\beta^{(3c)}_{\alpha\beta\gamma\delta}\rangle = F_{\mathrm{s}}\Delta^{+}_{\alpha\beta\gamma\delta} + F_{\mathrm{a}}\Delta^{-}_{\alpha\beta\gamma\delta}. \tag{3.77}$$

By solving the CDE (3.74) for nonlinear dipoles, we can express the factors F_{s} and F_{a} in terms of the products of the linear polarizabilities as follows [54]

$$F_{\mathrm{s}} = \frac{1}{15}Z^3Z^* f_{\mathrm{s}}\langle \mathrm{Tr}(\hat{\alpha}_i^{\mathrm{T}}\hat{\alpha}_i)\mathrm{Tr}(\hat{\alpha}_i^{\mathrm{T}}\hat{\alpha}_i^*) + 2\mathrm{Tr}(\hat{\alpha}_i^{\mathrm{T}}\hat{\alpha}_j\hat{\alpha}_i^{\mathrm{T}}\hat{\alpha}_i^*)\rangle,$$

$$F_{\mathrm{a}} = \frac{1}{6}Z^3Z^* f_{\mathrm{a}}\langle \mathrm{Tr}(\hat{\alpha}_i^{\mathrm{T}}\hat{\alpha}_i)\mathrm{Tr}(\hat{\alpha}_i^{\mathrm{T}}\hat{\alpha}_i^*) - \mathrm{Tr}(\hat{\alpha}_i^{\mathrm{T}}\hat{\alpha}_i\hat{\alpha}_i^{\mathrm{T}}\hat{\alpha}_i^*)\rangle. \tag{3.78}$$

Here $(\hat{\alpha}_j^{\mathrm{T}}\hat{\alpha}_j)_{\alpha\beta} \equiv \alpha_{j,\alpha'\alpha}\alpha_{j,\alpha'\beta}$, $(\hat{\alpha}_j^{\mathrm{T}}\hat{\alpha}_j^*)_{\gamma\delta} \equiv \alpha_{j,\beta'\gamma}\alpha^*_{j,\beta'\delta}$, and $(\hat{\alpha}_j^{\mathrm{T}}\hat{\alpha}_j\hat{\alpha}_j^{\mathrm{T}}\hat{\alpha}_j^*)_{\alpha\beta}$ $\equiv \alpha_{j,\alpha'\alpha}\alpha_{j,\alpha'\beta'}\alpha_{j,\gamma\beta'}\alpha^*_{j,\gamma\beta}$ (as above, the T symbol in $\hat{\alpha}^{\mathrm{T}}$ denotes a transposition of the matrix $\hat{\alpha}$). According to (3.77) and (3.78), the symmetry of the nonlinear polarizability of an isolated monomer [see (3.71)–(3.73)], is reproduced in the average polarizability of a cluster. The totally symmetric part of a monomer's nonlinear polarizability generates a totally symmetric part of the cluster polarizability ($F_{\mathrm{s}} \propto f_{\mathrm{s}}$); the same is true for the partially symmetrical parts ($F_{\mathrm{a}} \propto f_{\mathrm{a}}$).

The nonlinear susceptibility $\bar{\chi}^{(3c)}_{\alpha\beta\gamma\delta}$ responsible for DFWM in a composite material is defined via the relation

$$P^{(3c)}_{\alpha}(\omega) = 3\bar{\chi}^{(3c)}_{\alpha\beta\gamma\delta}(-\omega;\omega,\omega,-\omega)E^{(0)}_{\beta}E^{(0)}_{\gamma}E^{(0)*}_{\delta},$$

where $\bar{\chi}^{(3c)}_{\alpha\beta\gamma\delta}$ can be expressed in terms of the nonlinear polarizability averaged over an ensemble of clusters $\langle\beta^{(3c)}_{\alpha\beta\gamma\delta}\rangle$ as follows:

$$\bar{\chi}^{(3c)}_{\alpha\beta\gamma\delta}(-\omega;\omega,\omega,-\omega) = pv_0^{-1}\langle\beta^{(3c)}_{\alpha\beta\gamma\delta}(-\omega;\omega,\omega,-\omega)\rangle. \tag{3.79}$$

For the case when $\beta^{(3)}$ is due to the nonresonant electronic response (the low-frequency limit) we have $f_{\mathrm{a}} = 0$ [55]. Thus, the efficiency of four-wave mixing (which is proportional to the generated amplitude squared) is enhanced due to the clustering of particles in a composite material by the factor

$$G_{\mathrm{FWM}} = |F_{\mathrm{s}}/f_{\mathrm{s}}|^2 = \frac{(X^2+\delta^2)^4}{225} \times |\langle \mathrm{Tr}(\hat{\alpha}_i^{\mathrm{T}}\hat{\alpha}_i)\mathrm{Tr}(\hat{\alpha}_i^{\mathrm{T}}\hat{\alpha}_i^*) + 2\mathrm{Tr}(\hat{\alpha}_i^{\mathrm{T}}\hat{\alpha}_i\hat{\alpha}_i^{\mathrm{T}}\hat{\alpha}_i^*)\rangle|^2. \tag{3.80}$$

According to (3.80), the enhancement resulting from the particle clustering can be expressed via the product of linear polarizabilities α_i (averaged over an ensemble of clusters). These polarizabilities in turn represent the ratio of the local fields and the applied field, so that (cf. (2.2))

$$G_{\mathrm{FWM}} \sim \left| \frac{\left\langle \left| \mathbf{E}(\mathbf{r}) \right|^2 \left[\mathbf{E}(\mathbf{r}) \right]^2 \right\rangle}{\left[\mathbf{E}^{(0)} \right]^4} \right|^2 .$$

We can easily generalize (3.80) for the case of a nondegenerate FWM, such as CARS, so that we obtain

$$G_{\mathrm{FWM}} = |F_{\mathrm{s}}/f_{\mathrm{s}}|^2 = \frac{(X^2+\delta^2)^4}{225} |\langle \mathrm{Tr}[\hat{\alpha}_i^{\mathrm{T}}(\omega_s)\hat{\alpha}_i(\omega_1)] \mathrm{Tr}[\hat{\alpha}_i^{\mathrm{T}}(\omega_1)\hat{\alpha}_i^*(\omega_2)] + 2\mathrm{Tr}[\hat{\alpha}_i(\omega_s)^{\mathrm{T}}\hat{\alpha}_i(\omega_1)\hat{\alpha}_i^{\mathrm{T}}(\omega_1)\hat{\alpha}_i^*(\omega_2)]\rangle|^2 .$$

For a nonresonant excitation, when $|X(\omega)| \gg |w_n|$ and therefore $\alpha_i \approx \alpha_0$, we see from (3.80) that $G = 1$, i.e. there is no enhancement in this case.

We consider now expression (3.80) in more detail and introduce the quantity g_{FWM} defined as

$$g_{\mathrm{FWM}} \equiv \mathrm{Tr}(\hat{\alpha}_i^{\mathrm{T}}\hat{\alpha}_i)\mathrm{Tr}(\hat{\alpha}_i^{\mathrm{T}}\hat{\alpha}_i^*) + 2\mathrm{Tr}(\hat{\alpha}_i^{\mathrm{T}}\hat{\alpha}_i\hat{\alpha}_i^{\mathrm{T}}\hat{\alpha}_i^*). \tag{3.81}$$

Using (3.34), this expression can be transformed to

$$g_{\mathrm{FWM}} = \sum_{nmlk} [M_{nmlk} + 2K_{nmlk}] \Lambda_n \Lambda_m \Lambda_l \Lambda_k^*, \tag{3.82}$$

where, as above, $\Lambda_n \equiv [(w_n - X) - \mathrm{i}\delta]^{-1}$,

$$M_{nmlk} \equiv \sum_{jj'j''j'''} (i\alpha|n)(n|j\beta)(i\alpha|m)(m|j'\beta) \times (i\alpha'|l)(l|j''\beta')(i\alpha'|k)(k|j'''\beta'), \tag{3.83}$$

and K_{nmlk} is given by (3.59). Function M_{nmlk} satisfies the sum rule $\sum_{n,m,l,k} M_{nmlk} = 9$.

According to (3.80) and (3.81), enhancement factor G_{FWM} has the form

$$G_{\mathrm{FWM}} = \frac{(X^2+\delta^2)^4}{225} |\langle g_{\mathrm{FWM}}(X)\rangle|^2 . \tag{3.84}$$

By integrating (3.82) over X, we find the following sum rule for function g_{FWM}

$$\int_{-\infty}^{\infty} g_{\mathrm{FWM}}(X)\mathrm{d}X = 2\pi\mathrm{i} \sum_{nmlk} [M_{nmlk} + 2K_{nmlk}] R_{nk} R_{mk} R_{lk}, \tag{3.85}$$

where R_{nm} is defined in (3.60).

The product of Λ factors in (3.82) can be rewritten thus:

$$\Lambda_n \Lambda_m \Lambda_l \Lambda_k^* = R_{lk}\{\Lambda_n \Lambda_m \Lambda_l - R_{mk}\Lambda_n\Lambda_m + R_{nk}R_{mk}(\Lambda_n - \Lambda_k^*)\}. \tag{3.86}$$

As above for Raman scattering, it is instructive to find first $g_{\mathrm{FWM}}^{\mathrm{r}}$ which is due to the "resonance" difference of the eigenmode frequencies in (3.82), i.e. to calculate the contribution of only those modes for which $|w_n - w_m| \ll \delta$. In this case, the R factors become very large. Then, retaining in (3.86) only the term with the highest power of R and using (3.34), we obtain from (3.59), (3.82), and (3.83) the following ensemble-average expression

$$\langle g^{\mathrm{r}}_{\mathrm{FWM}}(X)\rangle = -\frac{15}{4\delta^3}\mathrm{Im}\alpha(X). \tag{3.87}$$

The quantity $\langle g^{\mathrm{r}}_{\mathrm{FWM}}(X)\rangle$ satisfies the sum rule

$$\int_{-\infty}^{\infty} \langle g^{\mathrm{r}}_{\mathrm{FWM}}(X)\rangle \mathrm{d}X = -\frac{15\pi}{4\delta^3}. \tag{3.88}$$

Our conjecture is that the spectral dependence predicted by the "resonance" contribution is correct in the general case, so that the enhancement factor G_{FWM} is as follows:

$$G_{\mathrm{FWM}} = C_{\mathrm{FWM}}\frac{(X^2+\delta^2)^4}{\delta^6}\left[\mathrm{Im}\alpha(X)\right]^2, \tag{3.89}$$

where the pre-factor C_{FWM} is an adjustable parameter. Note that the result obtained is in agreement with the estimate of (3.49).

Figure 3.13 shows results of numerical calculations of the enhancement factor G_{FWM} in silver CCAs, as a function (a) of spectral parameter X and as a function (b) of wavelength λ. The simulations were performed using exact formula (3.80), for 12 different CCAs containing 1 000 particles each. The solid lines in Figs. 3.13a and 3.13b describe the results of calculations based on formula (3.89), with C_{FWM} found from the relation $G_{\mathrm{FWM}}\delta^6 = 1$ in its maxima occurring at $X \approx \pm 4$. The dashed line in Fig. 3.13a represents a power-law fit for the range $0.1 \leq |X| \leq 3$, with $\delta = 0.05$. The computed exponent (8.31 ± 1.00) is close to 8 (only the case of negative values of X is shown; results for positive X are qualitatively similar). This value for the exponent is not surprising, since, within the interval $|X| \leq 3$, the dependence of $\mathrm{Im}[\alpha(X)]$ on X (see Fig. 3.3a) in (3.89) is relatively weak in comparison with that due to the factor X^8 in (3.89).

We conclude that formula (3.89) is in good agreement with numerical simulations, in a wide range, from the visible to the mid-infrared. Note that for $\lambda > 10$ μm the resonance condition $\lambda \ll \lambda_\tau$ does not hold and therefore the theory, strictly speaking, cannot be applied (for silver, $\omega_\tau \equiv \Gamma \approx 0.021$ ev, i.e. $\lambda_\tau \approx 56$ μm).

As seen in Figs. 3.13a and 3.13b, the enhancement strongly increases toward larger values of $|X|$ (when $X < 0$) or, in other words, toward longer wavelengths, where enhancements for the local fields are stronger (cf. Fig. 3.6).

It also follows from Fig. 3.13a that the product $G_{\mathrm{FWM}}\delta^6$ remains, on average, the same for the two very different values of δ chosen, namely 0.01 and 0.05. This indicates that, in accordance with (3.89), the enhancement is proportional to the sixth power of the resonance quality factor, i.e. $G_{\mathrm{FWM}} \propto Q^6$ $(Q \sim \delta^{-1})$ and reaches huge values in the long-wavelength part of the spectrum where $X \approx X_0 = -4\pi/3$.

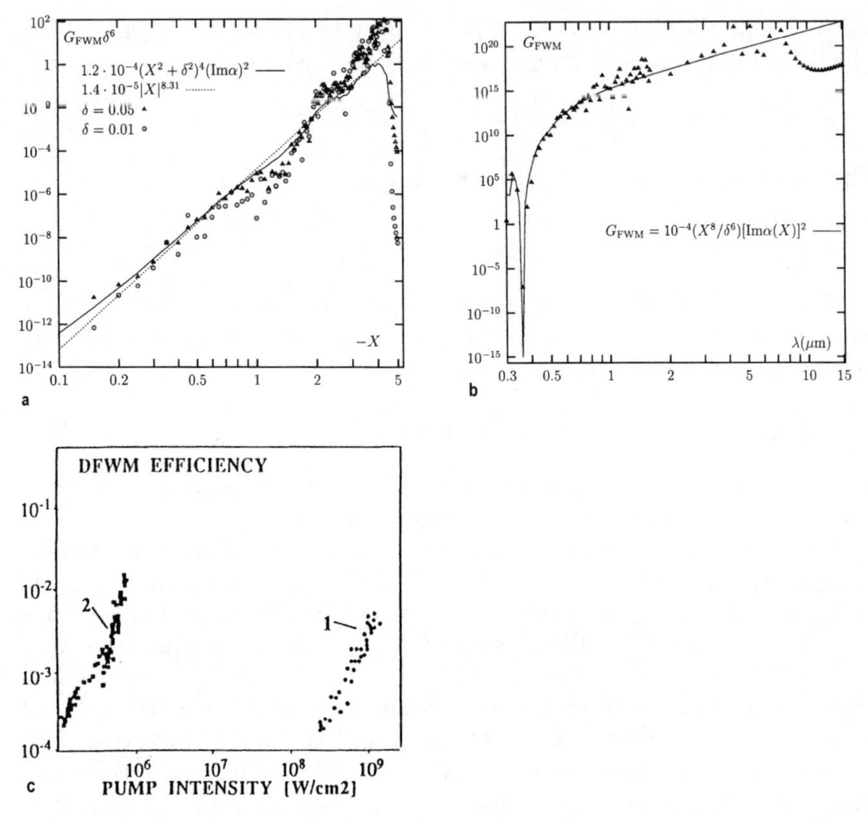

Fig. 3.13. Calculated enhancement of degenerate four-wave mixing, G_{FWM}: (**a**) $G_{FWM}\delta^6$ as a function of X $(X < 0)$ for CCAs; (**b**) G_{FWM} as a function of wavelength for silver CCAs. (**c**) Experimental results on DFWM efficiency *vs* pump intensity for silver particles which are isolated (1) and aggregated into fractal clusters (2) at $\lambda = 532\,\mathrm{nm}$

The nonlinear susceptibility $\bar{\chi}^{(3)}$ of the composite material, consisting of fractal aggregates of colloidal particles in some host medium (e.g. water) is given by $\bar{\chi}^{(3)} = p \cdot G_K \chi_m^{(3)}$ where $G_K \sim G_{FWM}^{1/2}$ is the enhancement of the Kerr optical nonlinearity (see also next section of this chapter), $\chi_m^{(3)}$ is the susceptibility of nonaggregated metal particles, and p is the volume fraction filled by metal. The experimentally measured value of $\chi_m^{(3)}$ [73, 74, 78] for silver monomers at $\lambda = 532\,\mathrm{nm}$ is $\chi_m^{(3)} \sim 10^{-8}\,\mathrm{esu}$. The value of $\chi_m^{(3)}$ reestimated for $\lambda = 532\,\mathrm{nm}$ from the data [96] is close to the above value. When the initially separated silver particles aggregate and fractal clusters are formed, a huge enhancement of the cubic susceptibility can occur [73].

A millionfold enhancement of degenerate FWM (DFWM) due to the clustering of initially isolated silver particles in colloidal solution was experimentally obtained [73]. Figure 3.13c shows the experimental data for conversion efficiency $\eta = I_s/I_1 \propto I_0^2$ (I_s, I_1 and I_0 are the intensities of the DFWM signal, probe beam and pump beam, respectively). As can been seen in Fig. 3.13c, similar values of η can be obtained in silver particles aggregated into fractal clusters at pump intensities $\sim 10^3$ times less than in the case of non-aggregated isolated particles. Since $\eta \propto I_0^2$, we conclude that the enhancement factor for silver fractal composites is $G \sim 10^6$. Note that in Fig. 3.13b calculations are done with the vacuum as a host medium, whereas the experimental data in Fig. 3.13c is obtained in a water colloidal solution. To compare experimental data with theory, we use Figs. 3.2 and 3.13a. As follows from Fig. 3.2, the values of X and δ for silver particles in water at laser wavelength $\lambda = 532$ nm are $X \approx -2.55$ and $\delta \approx 0.05$, respectively. According to Fig. 3.13a, $G_{\mathrm{FWM}} \sim 10^6$ to 10^7 for these values of X and δ, which is in reasonable agreement with the experimental observations.

The cubic susceptibility obtained experimentally for an aggregated sample is [74]: $|\bar{\chi}^{(3)}| = 5.7 \times 10^{-10}$ esu with $p \approx 5 \times 10^{-6}$. Note that p is a variable quantity and can be increased. We can assign the value 10^{-4} esu to the nonlinear susceptibility, $\chi^{(3f)}$, of the fractal clusters, i.e., $\bar{\chi}^{(3)} = p \cdot \chi^{(3f)}$, with $\chi^{(3f)} \sim 10^{-4}$ esu. This is a very large value for a third-order nonlinear susceptibility.

We also mention a high efficiency of four-wave mixing observed in films of J aggregates of pseudoisocyanine in a polymer matrix [74]. A 30-times increase in the nonlinear susceptibility of a film under doping of gold colloidal aggregates in a composite was detected. Very high optical susceptibilities, $|\bar{\chi}^{(3)}| \approx 10^{-6}$ esu, with a subpicosecond response time, were also measured.

Composites of particles with high intrinsic nonlinearities are of a particular interest for nonlinear optics and various applications, and metal fractal aggregates that can provide a significant enhancement of the local fields and nonlinearities fall into this category.

Finally in this subsection, we note that very large enhancement of high harmonic generation in fractal aggregates was predicted in [54]. Similar to other optical effects considered above, this enhancement strongly increases toward the infrared part of the spectrum, where the local fields associated with the fractal modes are more localized, on average, and much larger in amplitude. We note, however, that enhancement for harmonic generation is typically less than for the FWM process considered above that includes subtraction of photons of the applied fields in a four-photon process of generation of a nonlinear signal. Therefore, FWM is enhanced much more strongly than third-harmonic generation (THG) despite the fact that both processes are due to a nonlinearity of the same order, $\chi^{(3)}$ [54].

The reason is two-fold. First, in FWM there is additional enhancement of the nonlinear field amplitude since the generated frequency can also efficiently excite the fractal eigenmodes; for THG, the generated frequency is typically in the near UV spectral range where plasmon fractal modes do not provide much enhancement. Second, in processes that are characterized by a nonlinear susceptibility, with all frequencies of applied fields (say, ω_1, ω_2, and ω_3) entering the nonlinear susceptibility with a positive sign (so that generated frequency is $\omega_s = \omega_1 + \omega_2 + \omega_3$), the destructive interference between the nonlinear field amplitudes in different points of a random system results in a decreased overall enhancement [45]. In general, nonlinear processes involving subtraction of at least one of the frequencies (such as $\omega_s = \omega_1 - \omega_2 + \omega_3$) of incident waves are enhanced more significantly (see also Chap. 5).

3.6.4 Optical Kerr Effect

We now consider enhancement of Kerr optical nonlinearity. The Kerr polarizability has, in general, the form $\beta^{(3)}_{\alpha\beta\gamma\delta}(-\omega; \omega, \omega, -\omega)$ and it determines the nonlinear contribution (proportional to the field intensity) to the refractive index and absorption. The Kerr-type nonlinearity can also result in degenerate four-wave mixing (DFWM) considered above. Composite materials with large values of Kerr nonlinearity can be used as nonlinear optical filters and switches, which is important for applications in optical communication systems. Under certain conditions, they also demonstrate optical bistability, which can be utilized to build an optical analog of an electric transistor. Therefore, there is significant interest in developing materials with a large Kerr nonlinearity.

We consider enhancement of Kerr susceptibility due to the clustering of small particles embedded in a linear host material. We also assume that the volume fraction p filled by particles is small, and that they are initially randomly distributed in a host. Since p is small, interactions between particles before aggregation can be neglected. The aggregation results in spatially separated random clusters of particles. The interactions between the light-induced dipoles on particles within a cluster lead to the formation of collective eigenmodes; their resonant excitation results eventually in high local fields and the enhanced Kerr susceptibility.

The Kerr nonlinear polarizability $\beta^{(3)}$ has the same structure (see (3.70)) as the one describing DFWM. (In the present case, however, we assume that there is only one applied field, $\mathbf{E}^{(0)}$) The above consideration of DFWM is general, and most of the obtained results are applicable for other phenomena associated with the Kerr susceptibility.

For isotropic media, the Kerr polarizability can be written in the form of (3.71), with two independent constants, f_{s} and f_{a}. The polarization of the clusterized composite is then

$$P^{(3c)}_{\alpha}(\omega) = 3\bar{\chi}^{(3c)}_{\alpha\beta\gamma\delta}(-\omega; \omega, -\omega, \omega) E^{(0)}_{\beta} E^{(0)*}_{\gamma} E^{(0)}_{\delta} .$$

The effective Kerr susceptibility $\bar{\chi}^{(3c)}$ of the composite has the form:

$$\bar{\chi}^{(3c)}_{\alpha\beta\gamma\delta} = G_{\mathrm{K,s}} p \phi_{\mathrm{s}} \Delta^{+}_{\alpha\beta\gamma\delta} + G_{\mathrm{K,a}} p \phi_{\mathrm{a}} \Delta^{-}_{\alpha\beta\gamma\delta}, \tag{3.90}$$

where $\phi_{\mathrm{s,a}} = v_0^{-1} f_{\mathrm{s,a}}$, with v_0 being the volume of a particle, and

$$G_{\mathrm{K,s}} = \frac{1}{15} Z^3 Z^* \langle \mathrm{Tr}(\hat{\alpha}_i^{\mathrm{T}} \hat{\alpha}_i) \mathrm{Tr}(\hat{\alpha}_i^{\mathrm{T}} \hat{\alpha}_i^*) + 2\mathrm{Tr}(\hat{\alpha}_i^{\mathrm{T}} \hat{\alpha}_i \hat{\alpha}_i^{\mathrm{T}} \hat{\alpha}_i^*) \rangle \tag{3.91}$$

and

$$G_{\mathrm{K,a}} = \frac{1}{6} Z^3 Z^* \langle \mathrm{Tr}(\hat{\alpha}_i^{\mathrm{T}} \hat{\alpha}_i) \mathrm{Tr}(\hat{\alpha}_i^{\mathrm{T}} \hat{\alpha}_i^*) - \mathrm{Tr}(\hat{\alpha}_i^{\mathrm{T}} \hat{\alpha}_i \hat{\alpha}_i^{\mathrm{T}} \hat{\alpha}_i^*) \rangle . \tag{3.92}$$

The factors G_{s} and G_{a} are identical to $F_{\mathrm{s}}/f_{\mathrm{s}}$ and $F_{\mathrm{a}}/f_{\mathrm{a}}$ respectively (see (3.77) and (3.78)), and the enhancement factor for the DFWM process can be expressed in terms of $G_{\mathrm{K,s}}$ as $G_{\mathrm{FWM}} = |G_{\mathrm{K,s}}|^2$ (cf. (2.2)).

In general, according to (3.90)–(3.92), there are two different enhancement coefficients for totally symmetric ($\propto \Delta^+_{\alpha\beta\gamma\delta}$) and partially symmetric ($\propto \Delta^-_{\alpha\beta\gamma\delta}$) parts of the susceptibility in an isotropic system. The fact that there are two different independent constants for the Kerr response in an isotropic medium, results in particular in rotation of the polarization ellipse [55]. If the field $\mathbf{E}^{(0)}$ is polarized linearly or circularly, the nonlinear polarization $\mathbf{P}^{(3c)}$ can be expressed in terms of a single independent constant [55], F_{s} and $(F_{\mathrm{s}} + F_{\mathrm{a}})$ respectively. Also, in the low-frequency limit (where $\beta^{(3)}$ is due to the non-resonant electron response), the nonlinear susceptibility tensor must be fully symmetrical, i.e. $F_{\mathrm{a}} = 0$ for an arbitrary light polarization [55].

We now consider the enhancement associated with $G_{\mathrm{K,s}} \equiv G_{\mathrm{K}}$. The enhancement factor is, in general, complex: $G_{\mathrm{K}} \equiv G'_{\mathrm{K}} + \mathrm{i}G''_{\mathrm{K}}$. If $\beta^{(3)}$ is real, the real part G'_{K} and the imaginary part G''_{K} determine enhancement of the nonlinear refraction and nonlinear absorption, respectively.

In accordance with (3.89), we assume that G'_{K} is significantly larger than G''_{K} (an assumption that is supported by the numerical simulations of [54]) and that G'_{K} can be approximated thus:

$$G'_{\mathrm{K}} = C_{\mathrm{K}} \frac{X^4}{\delta^3} \mathrm{Im}\alpha(X), \tag{3.93}$$

where C_{K} is an adjustable parameter. Note that this formula is in agreement with the estimate (3.48).

In Fig. 3.14, plots of G'_{K} as a function of X (a) and of wavelength (b) are presented. The numerical calculations were performed using (3.91). The solid line in Fig. 3.14a represents the calculations based on (3.93), with the C_{K} chosen to satisfy the relation $|G'_{\mathrm{K}}\delta^3| = 1$ at its maximum at $X \approx -4$. From Fig. 3.14, we conclude that (3.93) approximates the numerical results reasonably well, from the visible to the mid-infrared parts of the spectrum (for wavelengths longer than $\lambda \sim 10\,\mu\mathrm{m}$ the resonance condition $\lambda \ll \lambda_\tau$ does not hold).

The Kerr-enhanced effect is proportional to the quality-factor cubed, i.e. to $Q^3 \sim \delta^{-3}$, and the following estimate is valid in the maximum: $G'_{\mathrm{K}}\delta^3 \sim 1$. Note that for most of frequencies G'_{K} is negative. The calculations for silver CCAs shown in Fig. 3.14b indicate that the Kerr optical effect can be enhanced in fractals by many orders of magnitude, and the enhancement

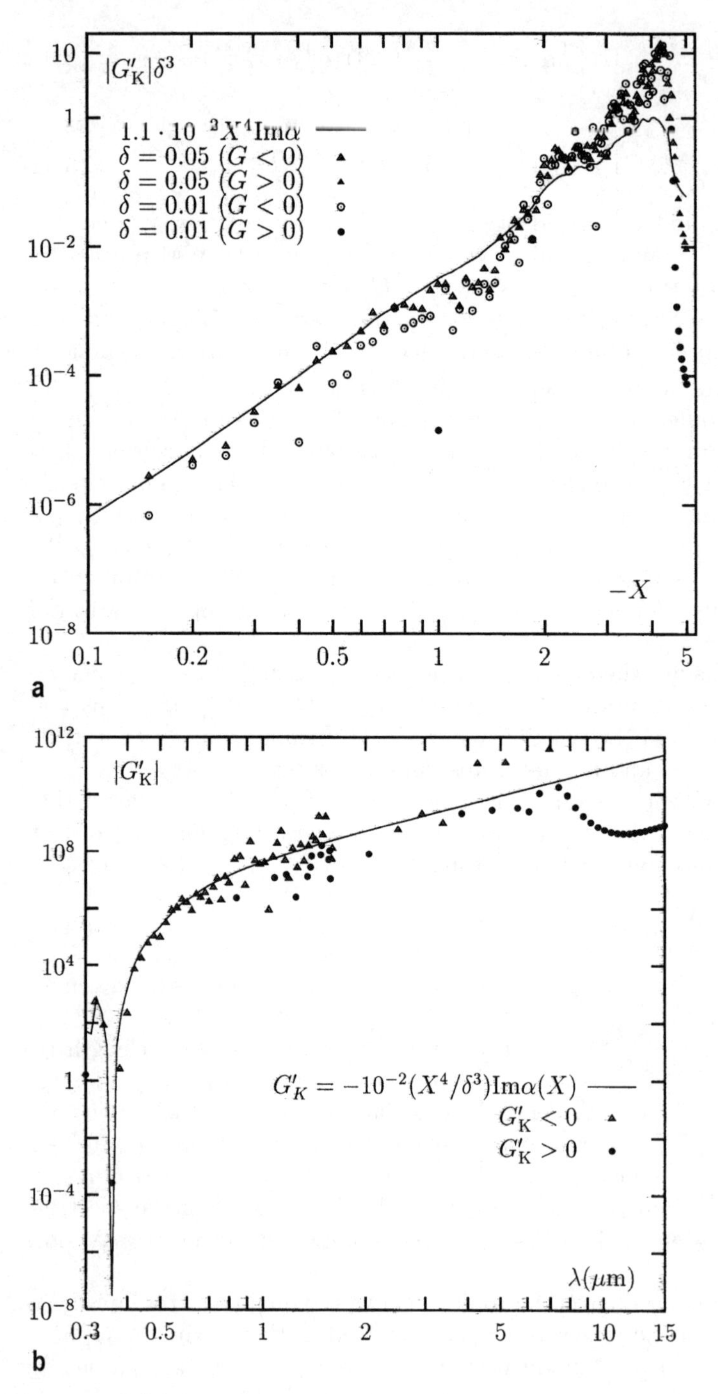

Fig. 3.14. (**a**) Enhancement of the real part of the Kerr optical susceptibility, G'_K. (**a**) $G'_K\delta^3$ as a function of X $(X < 0)$ in CCAs; (**b**) G'_K as a function of wavelength for silver CCAs

increases toward the infrared part of the spectrum. Calculations performed in [54] show that G''_K can also be large (although less thanG'_K), but it strongly fluctuates and changes its sign with frequency vary rapidly.

A very large enhancement for G_K indicates that optical materials based on composites consisting of small-particle clusters possess a high potential for various applications based on a large Kerr optical susceptibility.

To measure corrections to both nonlinear refraction and nonlinear absorption independently, it is possible to use the Z-scan technique [105]. Such measurements for fractal silver colloids were done in [74]. It has been found that, at $\lambda = 540\,\text{nm}$ and $p = 5 \times 10^{-6}$, the aggregation of silver colloidal particles into fractal clusters is accompanied by the increase of the nonlinear correction α_2 to the absorption, $\alpha(\omega) = \alpha_0 + \alpha_2 I$, from $\alpha_2 = -9 \times 10^{-10}\text{cm/W}$ to $\alpha_2 = -5 \times 10^{-7}\text{cm/W}$, i.e. the enhancement factor is $\sim 10^3$ [74]. (I is the laser intensity.) The measured nonlinear refraction at $\lambda = 540\,\text{nm}$ for fractal aggregates of silver colloidal particles is $n(\omega) = n_0 + n_2 I$, with $n_2 = 2.3 \times 10^{-12}\text{cm}^2/\text{W}$. Similar measurements at $\lambda = 1079\,\text{nm}$ give the values $n_2 = -0.8 \times 10^{-12}\text{cm}^2/\text{W}$ and $\alpha_2 = -0.7 \times 10^{-7}\text{cm/W}$.

The measured n_2 and α_2 allow determination of the real and imaginary parts of Kerr susceptibility [55, 105]; they are $\text{Re}[\bar{\chi}^{(3)}] = 1 \times 10^{-10}$ esu and $\text{Im}[\bar{\chi}^{(3)}] = -0.8 \times 10^{-10}$ esu for $\lambda = 540\,\text{nm}$, and $\text{Re}[\bar{\chi}^{(3)}] = -3.5 \times 10^{-11}$ esu and $\text{Im}[\bar{\chi}^{(3)}] = -2.7 \times 10^{-11}$ esu for $\lambda = 1079\,\text{nm}$. This means that the saturation of absorption and the nonlinear refraction provide comparable contributions to the nonlinearity. Note that the real part changes its sign with the wavelength. The measured enhancement factors are comparable with the calculated values of G_K, with accuracy of approximately one order of magnitude.

Using a different technique based on a dispersion interferometer, nonlinear correction n_2 to the refractive index was also measured at $\lambda = 1064\,\text{nm}$ and $p = 5 \times 10^{-6}$. The value obtained is $n_2 = -1.5 \times 10^{-11}\,\text{cm}^2/\text{W}$, which corresponds to $\text{Re}[\bar{\chi}^{(3)}] \approx -7 \times 10^{-10}$ esu [74].

Note that fractal aggregates of colloidal particles can be placed into a polymer matrix (like a gel). Then, thin films can be prepared with fractal aggregates in such matrices. The volume fraction filled by metal fractal aggregates in such thin films is typically larger than in the case of a colloidal solution, and therefore the nonlinearities are significantly higher [74].

A laser pulse duration used in the above experiments was $\sim 10\,\text{ns}$. The Kerr-type third-order nonlinearity was also detected with the use of 30 ps laser pulses [78]; however, the optical nonlinearities obtained were in this case smaller than in the experiments with nanosecond laser pulses. These studies indicate that there are probably two different types of optical nonlinearities $\bar{\chi}^{(3)}$. The smaller one has a time for nonlinear response in the picosecond (or even femtosecond) scale and the larger one in the nanosecond scale. The first of them can be associated with thermalization of the photoexcited "hot" electron gas in metal particles through electron-phonon coupling [96], whereas the second one probably involves effects connected with laser-induced heating of a crystal lattice of the metal.

When characterizing potential applications of materials, it is important to have large $\chi^{(3)}$ at a relatively small absorption. As a characteristic of the materials, the figure of merit F can be used that is defined via the ratio of the nonlinear susceptibility $\bar{\chi}^{(3)}$ of a composite material and its linear losses that are given by the imaginary part of the effective (linear) dielectric function, $\bar{\epsilon}'' = 4\pi \mathrm{Im}[\bar{\chi}^{(1)}]$. Thus, the figure of merit is defined through the relation $|\bar{\chi}^{(3)}/\bar{\epsilon}''| = F\chi_0^{(3)}$, where $\chi_0^{(3)}$ is a "seed" optical nonlinearity that can be either due to particles forming the composite or due to some nonlinear adsorbant molecules (we assume the former to be the case). Results of calculations of F for fractal silver aggregates (CCAs) are shown in Fig. 3.15, as a function of wavelength (in nm). It can be seen that the figure of merit in the fractals is very large, $\sim 10^7$, in the near infrared part of the spectrum where absorption is relatively small, whereas enhancement of optical nonlinearities is very significant (compare Fig. 3.5 with Figs. 3.13 and 3.14).

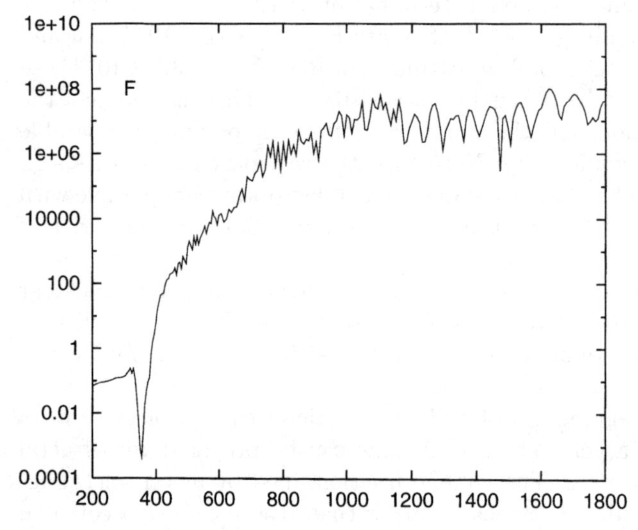

Fig. 3.15. The figure of merit F for silver fractal CCAs characterizing the ratio of nonlinear susceptibility $|\bar{\chi}^{(3)}|$ and linear losses $\bar{\epsilon}'' = 4\pi \mathrm{Im}[\bar{\chi}^{(1)}]$ (*see text*). The wavelength λ is given in nanometers

3.7 Local Photomodification in Fractals

We now consider experimental observations that support theoretical predictions of localization of optical excitations in fractals and frequency and polarization dependence of spatial locations of light-induced dipole modes, namely the hot spots previously described.

If a laser field creates hot spots in a particle aggregate, then the corresponding parts of the aggregate can be optically modified by a sufficiently powerful laser beam. As a result, the absorption corresponding to these parts will disappear, and there will be spectral holes left in the absorption for a given frequency and polarization. As shown in [50], this type of light-induced modification (selective in frequency and polarization) occurs at high laser intensities and results in local restructuring of resonance domains (hot spots) after irradiation of the cluster. The restructuring was attributed in [50] to sintering (coalescence) of particles, which can be observed when the laser pulse energy W per unit area is higher than a certain threshold W_{th}.

Electron micrographs of colloidal silver aggregates before [(a) and (c)] and after [(b) and (d)] irradiation by a sequence of laser pulses at two different wavelengths are shown in Fig. 3.16 [50]. Note that CCAs shown in Figs. 3.16a and 3.16c were prepared using two different methods, as described in [106] and [107], respectively. Figures 3.16b and 3.16d show the same two clusters but after photomodification. Comparison of the micrographs of the cluster before and after irradiation at laser wavelength $\lambda_{\mathrm{L}} = 1079$ nm (Figs. 3.16a and 3.16b, respectively) shows that the structure of the cluster as a whole remains the same after the irradiation, but monomers within small nm-sized domains change their size, shape, and local arrangement. The minimum number of monomers in the region of modification is 2-3 at $\lambda_{\mathrm{L}} = 1079$ nm. Thus, the resonance domain at $\lambda_{\mathrm{L}} = 1079$ nm can be as small as $\lambda_{\mathrm{L}}/25$.

Although there are fluctuations in both shape and size of the modified domains, Fig. 3.16b reveals that hot zones associated with resonant excitation are highly localized, in accordance with theoretical predictions. When a laser wavelength is close to the monomer absorption peak, $\lambda_{\mathrm{L}} \sim 450$ nm (Figs. 3.16c and 3.16d), localization of optical excitations is much weaker. We estimate that about 70% of all monomers were photomodified at $\lambda_{\mathrm{L}} = 450$ nm (see Fig. 3.16d), while only about 10% of monomers were modified at $\lambda_{\mathrm{L}} = 1079$ nm. Note that the ratio W/W_{th} was approximately the same in both cases. The increase of localization of optical excitations in fractals toward longer (with respect to the monomer absorption peak) wavelengths was predicted theoretically in [46].

Photomodification leads, in turn, to a spectral hole in the aggregate absorption spectrum in the vicinity of the laser wavelength [50, 79]. An example of spectral hole burning is shown in Fig. 3.17. The dependence of the hole depth on laser intensity demonstrates a threshold characteristic of the photomodification process [50, 79].

In accordance with the increase of localization of optical excitations toward longer wavelengths, the threshold energy measured in [50] also decreases toward the infrared. This is because photomodification occurs primarily for most absorbing monomers, i.e. those that are in the hot spots; toward longer

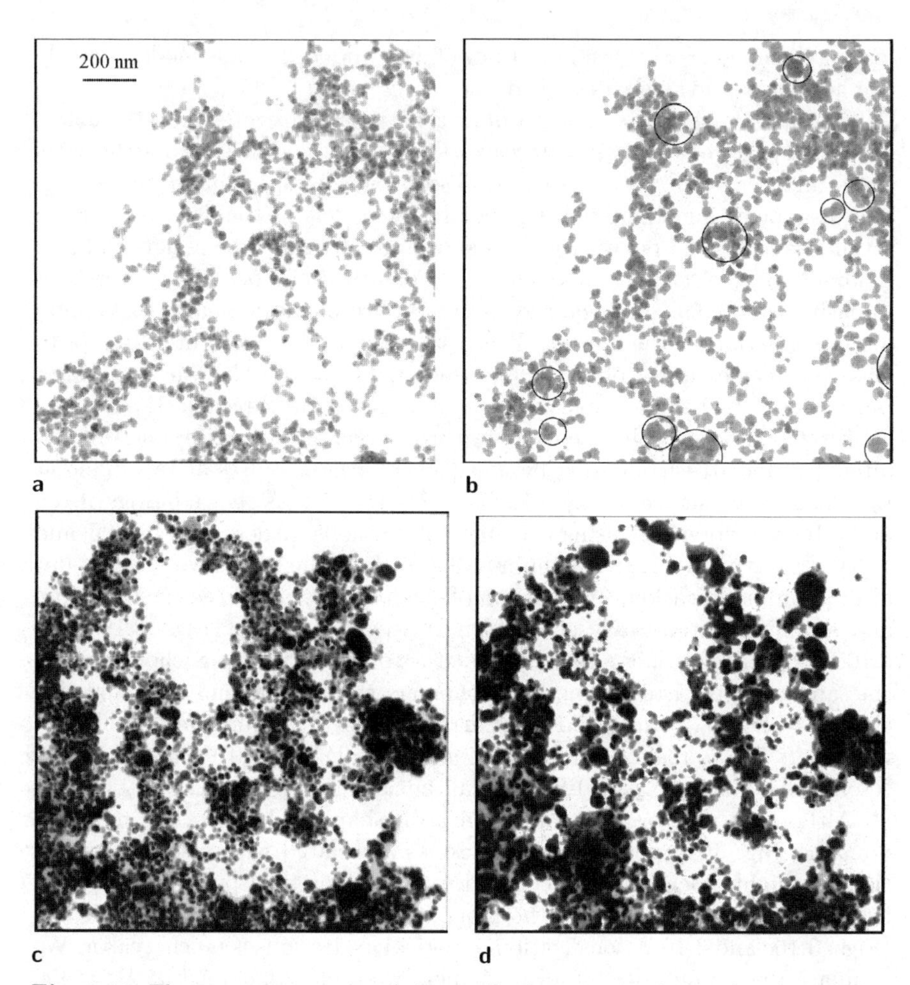

Fig. 3.16. Electron micrographs of silver colloid aggregates before (**a, c**) and after (**b, d**) irradiation by laser pulses at wavelength $\lambda_{\mathrm{L}} = 1079$ nm and energy density per unit area $W = 11$ mJ/cm^2 (**b**), and $\lambda_{\mathrm{L}} = 450$ nm and $W = 20$ mJ/cm^2 (**d**). The circles in (**b**) are merely to aid the eye

wavelengths, the size of the hot spots and the average number of monomers in them both decrease, leading to lower thresholds for the photoinduced particle restructuring. This effect is illustrated in Fig. 3.18, where the spectral dependence of the threshold absorbed energy resulting in photomodification is shown as a function of wavelength. Specifically, Fig. 3.18 shows the energy Q absorbed per unit volume in the layer where photomodification occurs, with $Q(\lambda_{\mathrm{L}}) = \alpha(\lambda_{\mathrm{L}})W_{\mathrm{th}}(\Lambda_{\mathrm{L}})$. As seen in Fig. 3.18, the threshold-absorbed energy significantly decreases as λ_{L} changes from 355 nm to 2000 nm, in agreement with theoretical calculations [50].

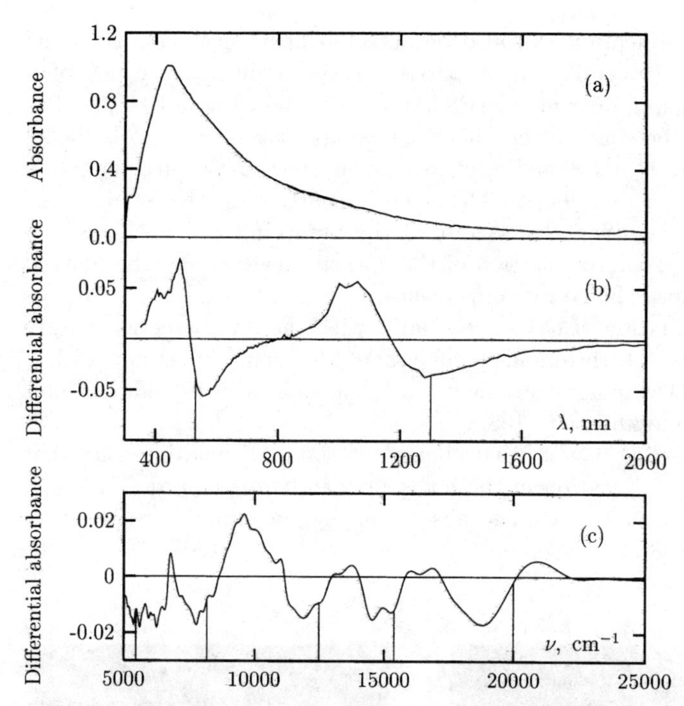

Fig. 3.17. (**a**) Absorption spectrum of silver aggregates in a gelatin film. (**b**) Difference in the absorbance of the silver-gelatin film before and after irradiation by laser pulses at $\lambda_L = 532\,\mathrm{nm}$, $W = 24\,\mathrm{mJ/cm^2}$ and at $\lambda_L = 1300\,\mathrm{nm}$, $W = 11\,\mathrm{mJ/cm^2}$. (**c**) An example of five spectral holes recorded at the same area. The vertical bars denote laser frequencies. W/W_{th}=1.1–1.3; the number of pulses increases from 5 in the visible spectral range to 30 in the infrared range in order to obtain the holes of the same depth

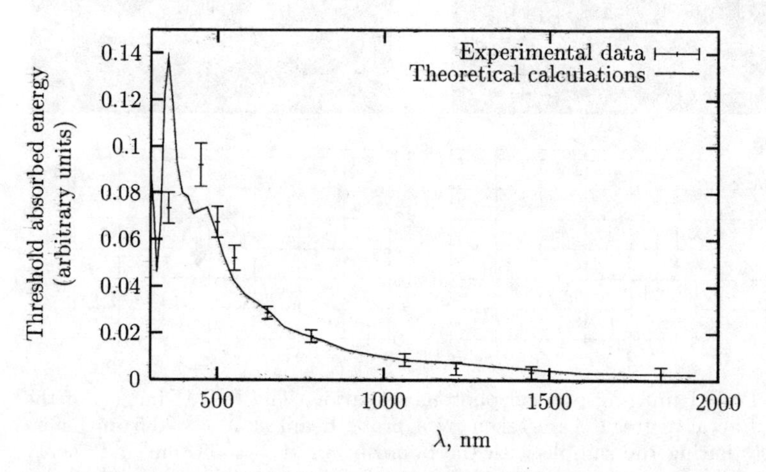

Fig. 3.18. Spectral dependence of the threshold absorbed energy per unit volume, αW_{th}

Local photomodification can also be detected using near-field scanning optical microscopy (NSOM) and, in particular, its modification called photon scanning tunneling microscopy (PSTM), as shown in Fig. 3.19 [108]. The PSTM images of the same silver fractal aggregate, taken by a probe beam at $\lambda = 543.5\,\mathrm{nm}$ both before and after irradiating the sample with the pump at $\lambda = 532\,\mathrm{nm}$ and energy density $9.5\,\mathrm{mJ/cm^2}$, clearly show that some of the hot spots degrade significantly because of the pump-induced local restructuring (sintering). The cross-section of the optical images along the marked horizontal line is also shown for convenience.

Note that irradiation of the same sample with the pump energy density only slightly below the threshold, namely at 8.5 $\mathrm{mJ/cm^2}$, does not lead to any changes in PSTM images; this fact clearly indicates a threshold character of the local photomodification [108].

It is also interesting to note that changes in the local field intensity that result from photomodification in the hot spots can be of the order of 100%, whereas the change in the average absorption by the sample is negligible, being below 1% [108].

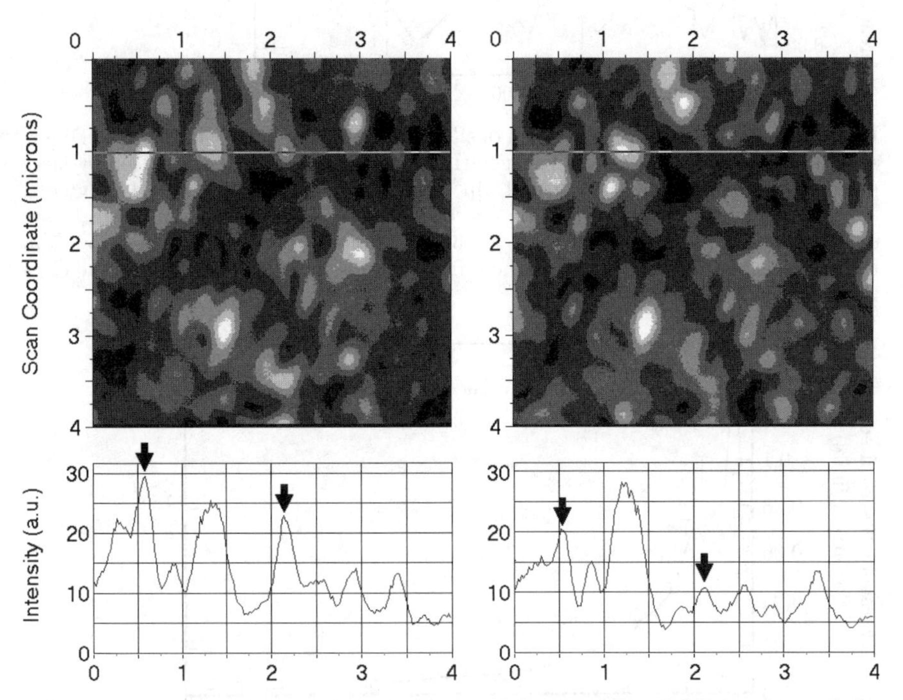

Fig. 3.19. PSTM imaging of local photomodification. The PSTM images of the same silver fractal aggregate are taken by a probe beam at $\lambda = 543.5\,\mathrm{nm}$ before and after irradiating the sample with the pump beam at $\lambda = 532\,\mathrm{nm}$ and energy density $9.5\,\mathrm{mJ/cm^2}$. The cross-section of the optical images along the marked line is also shown for convenience

3.8 Fractals in Microcavities

As shown in this section above, optical excitations in a fractal composite may be localized in regions significantly smaller than the wavelength, so that the corresponding electromagnetic energy becomes concentrated in regions smaller than the diffraction limit of conventional optics, resulting in large local fields and strong enhancement of optical nonlinearities. Seeding the aggregates into microcavities further increases the local fields because of light trapping by microcavity resonance modes. This class of optical materials, microcavities doped with nanostructured fractal aggregates, is anticipated to possess unique properties with high potential for application in laser physics, optoelectronics, and photonics.

The excitation of morphology-dependent resonances (MDRs) in dielectric microcavities allows achievement of large enhancement of the optical response [109]. These resonances, which may have extremely high quality factors ($Q = 10^5$ to 10^9), result from confinement of the radiation within the microcavity by total internal reflection. Light emitted or scattered in the microcavity may couple to the high-Q MDRs lying within its spectral bandwidth, leading to enhancement of both spontaneous and stimulated optical emissions.

Strong existing evidence suggests that fractal nanocomposites and microcavity resonators individually result in large enhancements of optical emissions. In [110] it was demonstrated that huge, multiplicative enhancement factors can be obtained under the simultaneous, combined action of these two resonant processes when the emitting species is adsorbed onto metal fractal aggregates contained within high-Q microcavities.

In these experiments, it was found that lasing emission from Rhodamine 6G (R6G) dye molecules adsorbed onto silver colloidal aggregates inside a cylindrical microcavity can be obtained for dye molarities approximately three orders of magnitude lower than for the corresponding microcavity dye laser in the absence of colloidal aggregates, and for a threshold pump intensity approximately three orders of magnitude less than for a conventional dye laser.

Also in [110], the emission intensity of different spectral components of the photoluminescence was studied as a function of the pump intensity. It was found that this dependence is linear for low-excitation intensities for all components. However, when the pump intensity exceeds some critical value in the range between 20-50 W/cm^2, some peaks grow dramatically, exhibiting a lasing threshold dependence (Figs. 3.20a and 3.20b). The threshold power for λ_L=543.5 nm HeNe laser excitation is as small as 2×10^{-4} W. It is noteworthy that the R6G concentration was only 5×10^{-7} M in these experiments, which as mentioned is three orders of magnitude lower than that for conventional dye lasers with an external cavity and three orders of magnitude lower than that for a microdroplet laser without silver fractal aggregates [111]. In [110] the minimum R6G concentration that results in lasing was as low as 10^{-8} M. These findings suggest that the lasing effect is due to dye molecules

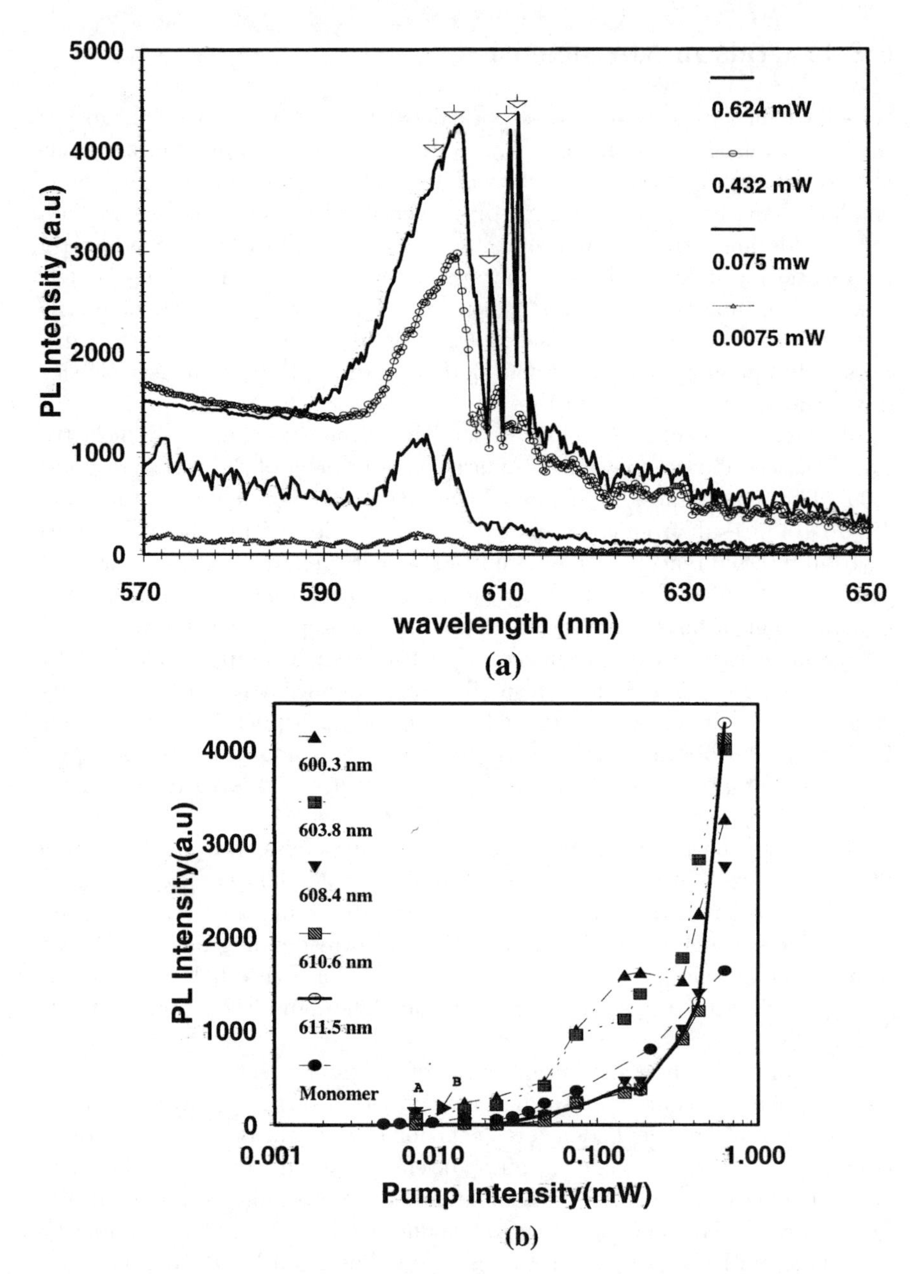

Fig. 3.20. (**a**) Photoluminescence spectra of 5×10^{-7} M R6G dye/fractal aggregate solution in a microcavity for $\lambda_L = 543.5$ nm HeNe laser excitation at different pump powers. (**b**) Lasing threshold effect for selected spectral components [arrows in (**a**)]. For comparison, the pump-power dependence of the luminescence spectrum of a dye-doped non-aggregated monomer solution at 611.5 nm is also shown; the intensity has been increased by a factor of 10^2, for which the intensities of points A and B are 5 and 20 units

adsorbed on the surface of fractal silver aggregates. This conclusion is also supported by the fact that increasing the R6G concentration to 10^{-5} M does not result in additional growth of the lasing peak intensities; the additional dye concentration is apparently not adsorbed onto the silver particles, but remains in solution as free molecules, where it does not effectively contribute to the enhanced lasing effect.

The experiments desribed in [110] show that placing R6G dye in a microcavity leads to enhancement of dye photoluminescence by a factor between 10^3 and 10^5. By adding nonaggregated silver colloidal particles to the dye solution in a microcavity, further multiplicative enhancement is obtained, varying between 10^2 and 10^3. Finally, aggregation of the initially isolated colloidal particles in the microcavity into fractals results in a final multiplicative enhancement factor that varies between 10^3 and 10^4. Combining these multiplicative enhancement factors, the overall emission enhancement provided by fractal/microcavity composite media can be enormously large, varying in the experiments in [110] between 10^8 and 10^{12}.

An extremely strong, time-dependent interaction was also observed in [110] when a dye solution, initially containing nonaggregated silver monomers, was irradiated by a cw pump laser. Figure 3.21 illustrates this for the case of 0.75 mW, $\lambda_L = 543.5$ nm HeNe pump light. The interaction is seen to exert a strong effect not only on the intensity of the peaks but also on their spectrum. In fact, inspection of Fig. 3.21 reveals that several peaks near 610 nm are driven above the lasing threshold during the irradiation process; however, a more precise statement is that the lasing threshold decreases with time for these peaks. It is known that irradiation of an initially nonaggregated colloidal solution promotes aggregation [112, 113]; this photostimulated aggregation has been conjectured as being induced by surface-enhanced ionization of the metal monomers [113]. Hence, a plausible explanation of the strong time-dependent interaction displayed in Fig. 3.21 is that, as time progresses so does the degree of aggregation, and when the aggregation reaches a particular level, fractal resonant enhancement lowers the lasing threshold to the level that is, in fact, being provided by the HeNe pump laser. Another plausible explanation of the time-dependent interaction can be based on light-induced pulling of fractals into the high-intensity area associated with the whispering gallery modes. In either case, the above experimental observation provides direct evidence of the effect of fractal media on lasing emission. It is also an important result for microtechnology, since, for example, it offers the possibility of activating lasing emission by remote, non-invasive techniques.

Microcavity surface-enhanced Raman scattering (SERS) was also investigated in [110]. SERS spectra from sodium citrate molecules adsorbed onto silver fractal aggregates in a microcavity were obtained in these experiments under conditions where MDRs either were, or were not, excited. It was found that SERS is 10^3 to 10^5 times more intense when MDRs are excited. However, of greater interest is the coupled, multiplicative enhancement factor caused

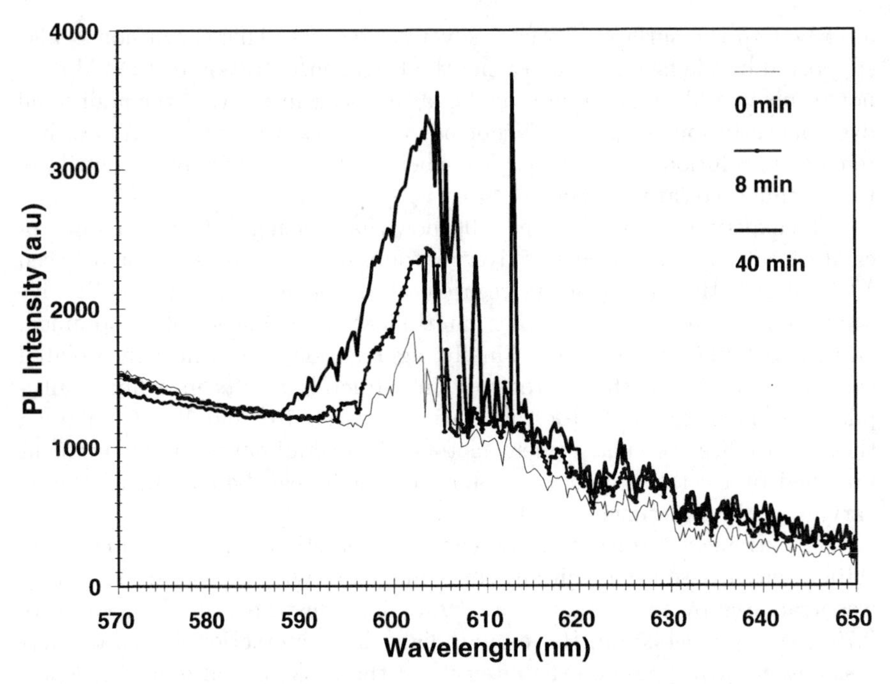

Fig. 3.21. Influence of photostimulated aggregation of colloidal particles for different irradiation times for $\lambda_L = 543.5$ nm HeNe laser excitation (maximum power, 0.75 mW); laser focused into a microcavity with a 35 μm spot size. Aggregation limited to a small volume; moving pump beam approximately 1 cm from previous irradiation region in sample returns the spectrum to a zero-time monomers-only trace

by both fractal aggregates and microcavities. By comparing Raman signal levels from sodium citrate adsorbed on silver colloid aggregates and from a high-concentration sodium citrate solution without colloidal particles, it was found that SERS enhancement resulting from fractal aggregation of colloidal silver is a factor of the order of 10^5 to 10^6, which is consistent with the data of [52]. Thus, with the additional, multiplicative enhancement provided by the microcavity, the resultant average SERS enhancement by fractals in a microcavity is estimated to be in the range 10^8 to 10^{11}.

Finally, it is important to mention that the local enhancement factors for optical emissions from fractal aggregate/microcavity composite media may be significantly greater than the average enhancements measured in the far-zone. It was recently shown, both theoretically and experimentally, that local SERS enhancements for metal nanoparticle fractal aggregates may be as high as factors of 10^{12} to 10^{15}, which makes possible observation of single-molecule Raman scattering [92–94]. However, since the enormous enhancements occur in the near-field zone and are localized within sub-wavelength hot spot regions of the fractal medium, these intense near-field electromagnetic amplitudes can be coupled to microcavity MDRs. In such a case, this would result in coupled, multiplicative enhancements in fractal/microcavity composite media that are truly enormous. For example, using 10^{12} as a representative fractal hot-spot enhancement factor, and taking 10^5 as an additional multiplicative microcavity MDR enhancement factor, we find that the local enhancement in the hot spots of fractals placed in a microcavity may be as large as 10^{17}. These potential advances, foreshadowed by the remarkably large enhancements actually observed in the study [110], open new fascinating possibilities for optical microanalysis, including, for example, Raman spectroscopy of individual molecules, and nonlinear optical studies of single molecules and nanoparticles.

We note that the observed multiplicative enhancement provided by microcavities and fractals opens a new feasibility to enhance the spontaneous emission rate (the Purcell effect), with the use of smaller microcavities, and development of high-efficiency light emitters (see also Sect. 3.4)

To summarize this section, we note that the above results promise an advance in the design of micro/nanolasers, operating on a small number of, or even on individual, molecules adsorbed on metal nanostructures within a microcavity, as well as offering the possibility of combining surface-enhanced radiative processes and high-Q morphology-dependent resonances in microcavities. Another exciting application can be based on doping optical fibers with fractals.

4. Self-Affine Thin Films

I don't feel frightened by not knowing things, by being lost in a mysterious universe ... I have approximate answers and possible beliefs and different degrees of certainty about different things, but I'm not absolutely sure of anything.

Richard Feynman

4.1 General Approach

We consider in this chapter the linear and nonlinear optical properties of self-affine thin films. Rough thin films, for instance, formed when an atomic beam condenses onto a low-temperature substrate, are typically self-affine fractal structures [10]. Contrary to the case of "usual" roughness, there is no correlation length for self-affine surfaces, which implies that inhomogeneities of all sizes are present, within a certain size interval, according to a power-law distribution. Self-affine surfaces obtained in the process of the film growth belong to the Kardar-Parisi-Zhang universality class.

Although self-affine structures differ from self-similar fractal objects (to reveal the scale-invariance they require two different scaling factors, in the surface plane and in the normal direction), the optical properties of self-affine thin films are, in many respects, similar to those of fractal aggregates considered in Chap. 3 [114–116]. For example, both fractal aggregates and self-affine films possess dipolar eigenmodes distributed over a wide spectral range [11, 114]. In contrast, for the case of conventional (non-fractal) random ensembles of monomers, such as the random gas of particles or randomly distributed close-packed spheres, the absorption spectra peak near a relatively narrow resonance of an individual particle, as illustrated in Fig. 3.3. In fractals, a variety of dipolar eigenmodes can be excited by a homogeneous electric field, whereas only one dipolar eigenmode can be excited in, for example, a small dielectric sphere. These striking differences are explained by localization of optical excitations in various random, spatially separated, parts of a fractal object, such as a self-affine surface [11, 46].

In random but homogeneous (on average) media, dipolar modes (polaritons) are, typically, delocalized over large spatial areas. All monomers absorb light energy with approximately equal rate in the regions whose linear dimensions significantly exceed the incident-field wavelength. This is, however, not the case for fractal nanocomposites and self-affine films. Optical excitations in these fractal objects tend to be localized. Because of this localization, and because there is a large variety of different local geometrical structures in fractals that resonate at different frequencies, the fractal optical modes cover a large spectral interval.

Most rough surfaces are self-affine within a certain interval of sizes and, therefore, their optical properties are typical for fractals. Because the field distributions are extremely inhomogeneous at the rough surfaces of thin films, there are "cold" regions of small local fields and "hot" areas of high local fields. Strong enhancement of a number of optical phenomena in rough metal films is associated with much higher values of local fields in the hot spots of a rough film.

To simulate a self-affine film we use the restricted solid-on-solid (RSS) model (for details see [117] and references therein). In this model, a particle is incorporated into the growing aggregate only if the newly created interface does not have steps that are higher than one lattice unit a. The surface structure of such deposits is relatively simple, because there are no overhangs. In this way, strong corrections to scaling effects are eliminated and the true scaling behavior appears clearly, even for small dimensions.

In the long term, the height-height correlation function for a self-affine surface has the form

$$\langle [h(\mathbf{r}) - h(\mathbf{r}+\mathbf{R})]^2 \rangle \sim R^{2H}, \tag{4.1}$$

where $\mathbf{R}$ is the radius-vector in the plane normal to the growth direction z and the scaling exponent (co-dimension) H is related to the fractal dimension D through the formula $H = 3 - D$. For the RSS model, $D = 2.6$ and the scaling formula above is valid for large values of the average height $\bar{h}$ (which is proportional to the deposition time), such that $\bar{h} \gg l^{\zeta}$, where $\zeta = 2(d+1)/(d+2) = 2 - H$; here l is the linear size of a system and d is the dimension of the embedding space. Simulations in [114–116] satisfied this condition, and the above scaling relation was well manifested.

In the simulations, the bulk (regular) part of the computer-generated film was removed so that the resultant sample had at least one hole. Clearly, the removal of the bulk of a film does not affect the scaling condition (4.1). A typical simulated self-affine film is shown in Fig. 4.1.

Unlike "conventional" random surfaces, contributions of higher spatial Fourier harmonics (with the harmonics' amplitudes larger than their wavelengths) plays an important role in the Fourier expansion of a self-affine surface profile. This means that neither the Rayleigh perturbation approximation [118–120] nor the Kirchhoff (geometrical optics) approach can be directly applied to describe optical properties of self-affine structures [121]. Apart from these two basic approaches, there exists a phase perturbation

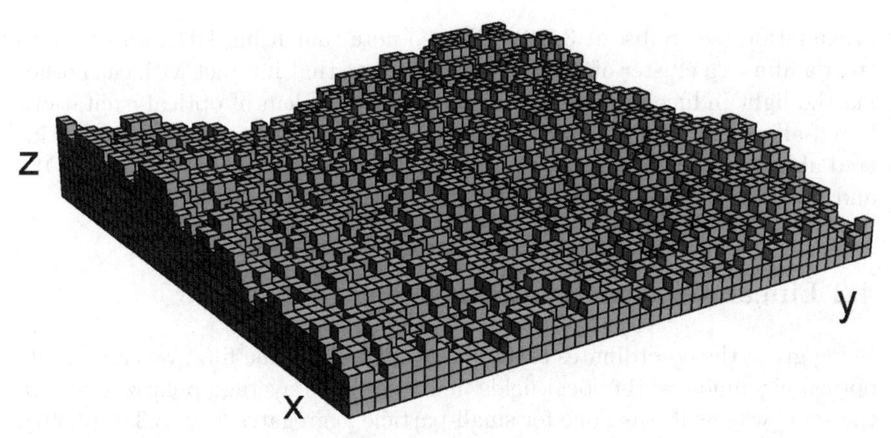

Fig. 4.1. Self-affine film obtained in the restricted solid-on-solid model

method [122, 123], which is an intermediate between the former two methods and also cannot be applied to self-affine surfaces.

We use approach based on discrete dipole approximation (DDA). DDA was originally suggested by Purcell and Pennypacker [81] and developed in later papers [82] to calculate optical responses from an object of an arbitrary shape. It is based on replacing an original dielectric medium by an array of point-like elementary dipoles. In the previous chapter we discussed how DDA can be also applied to fractal clusters built from a large number of small interacting monomers.

Following the main idea of DDA, we model a self-affine film by point dipoles placed according to an algorithm (described below) in sites of a simple cubic lattice with a period a, which is assumed to be much smaller than the size of any spatial inhomogeneities. The occupied sites correspond to the spatial regions filled by the film, while empty sites correspond to empty space. The linear polarizability of an elementary dipole (monomer) α_0 is given by the Lorentz-Lorenz formula, having the same form as the polarizability of a dielectric sphere with radius $R_{\mathrm{m}} = (3/4\pi)^{1/3}a$:

$$\alpha_0 = R_{\mathrm{m}}^3[(\epsilon - 1)/(\epsilon + 2)], \tag{4.2}$$

where as above, $\epsilon = \epsilon' + \mathrm{i}\epsilon''$ is the bulk dielectric permittivity of the film material. Note that (4.2) coincides with (3.27), if $\epsilon_{\mathrm{h}} = 1$. The choice of the sphere radius R_{m} provides equality of the cubic lattice elementary cell volume (a^3) to the volume of the effective sphere (monomer) that represents a point-like dipole ($4\pi R_{\mathrm{m}}^3/3$) [81, 82]. Consequently, for large films consisting of many elementary dipoles, the volume of the film is equal to the total volume of the imaginary spheres. Note that neighboring spheres intersect geometrically because $a < 2R_{\mathrm{m}}$.

Using the intersecting spheres allows one, to some extent, to take into account the effects of the multipolar interaction within the pure dipole approximation (see Subsect. 3.2.2). We also note that using DDA allows us to treat a film as a cluster of polarizable monomers that interact with each other via the light induced dipoles. This makes the problem of optical excitations in self-affine thin films similar to that in fractal aggregates, which was considered above in Chap. 3. Thus we can use the coupled-dipole equation (CDE) and the solutions to it, as discussed in Chap. 3.

4.2 Linear Optical Properties

Being given the co-ordinates of all dipoles in a self-affine film, we can find its optical eigenmodes, the local fields, and the film's average polarizability, in the same way as it was done for small-particle aggregates (see (3.33)–(3.39)). A reader interested in details of theoretical calculations of optical responses for a system of point dipoles should read Subsect. 3.2.4 of the previous chapter. Below in this chapter we discuss results of calculations of optical properties of self-affine films generated in the RSS model. These properties are found by applying first the DDA approximation and then solving CDE in the quasi-static approximation, which allows us to find the local fields and dipoles. Those in turn can be used to find the average optical characteristics, such as the film polarizability.

Figure 4.2 shows plots for the imaginary parts of the "parallel" and "perpendicular" components of the mean polarizability per particle, $\alpha_{\|} \equiv (1/2)\langle\alpha_{i,xx} + \alpha_{i,yy}\rangle$ and $\alpha_{\perp} \equiv \langle\alpha_{i,zz}\rangle$, describing absorption of the film at different light polarization. The "parallel" component $\alpha_{\|}$ characterizes the polarizability of a self-affine film in the (x, y) plane, whereas the "perpendicular" component $\alpha_{\perp}$ gives the polarizability in the normal z direction. The polarizability components satisfy the sum rule: $\int \alpha_{\perp,\|}(X)\mathrm{d}X = \pi$ (see (3.37)).

From Fig. 4.2 it is clear that there is strong dichroism expressed in the difference between the two spectra, $\alpha_{\|}(X)$ and $\alpha_{\perp}(X)$. The modes contributing most to $\alpha_{\|}$ (the "longitudinal" modes) are located in the long-wavelength part of the spectrum (negative X; cf. (3.30)), whereas the "transverse" modes tend to occupy the short-wavelength part of the spectrum (positive X) (see also Fig. 3.2a). To some extent, this can be understood by roughly considering a film as an oblate spheroid, where the longitudinal and transverse modes are shifted to red and blue respectively with respect to a resonance of a sphere. However, in contrast to the case of a spheroid, there is a large variety of eigenmodes in a self-affine film, as follows from Fig. 4.2. Really, the widths of the spectra in Fig. 4.2 are much larger than the width of an individual resonance, which is given by $\delta = 0.03$ (see (3.31)); this indicates a strongly inhomogeneous broadening associated with a variety of dipolar eigenmodes on a self-affine surface. Thus, the dipole-dipole interactions of constituent dipoles (monomers) in a self-affine film generate a wide spectral range of resonant modes, similar to the case of fractal aggregates.

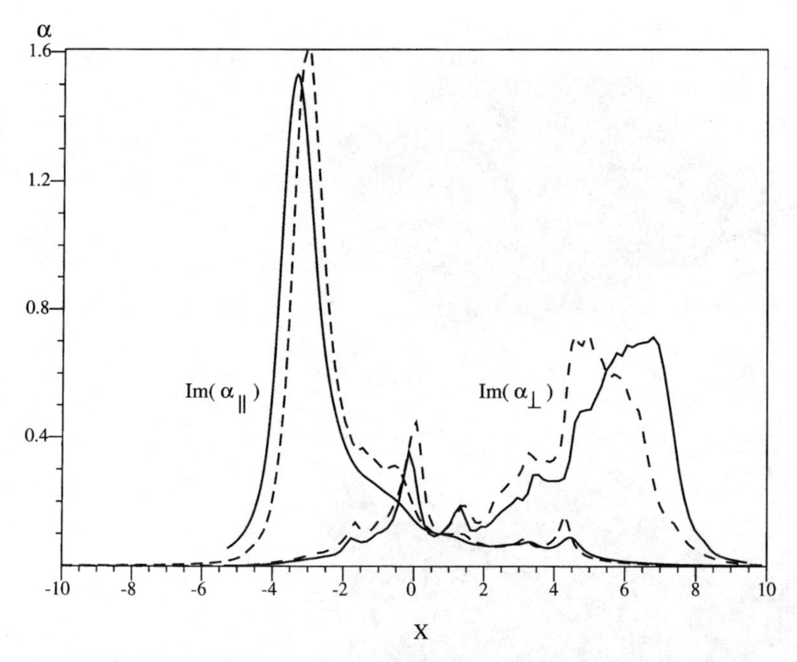

Fig. 4.2. Imaginary parts of the parallel, $\alpha_{\|}$, and perpendicular, $\alpha_{\perp}$, components of the polarizability (absorption). Results for samples with $N \sim 10^4$ (*solid lines*) and $N \sim 10^3$ (*dashed lines*) dipoles each are shown

From Fig. 4.2, we also make an important conclusion, that in the quasi-static approximation the optical properties of a self-affine film do not depend on the number of monomers N nor, therefore, on linear size l of a film. The calculations performed for ensembles of samples with very different numbers of particles and linear sizes give similar results. Note also that the fact that the spectra are almost independent of the number of dipoles N justifies the discrete dipole approximation that has been used.

The field distributions of eigenmodes on a self-affine surface are extremely inhomogeneous. On such a surface, there are hot spots associated with areas of high local fields and cold zones associated with small local fields. (A similar patchwork-like picture of the field distribution is observed in fractal clusters, as discussed in Chap. 3.) Spatial locations of the modes are very sensitive to both frequency and polarization of an applied field.

To demonstrate this, Fig. 4.3 shows the intensity distributions for local fields $|\mathbf{E}(\mathbf{R}_i)|^2$ on the film-air interface. (Note that $E_{i,\alpha} \equiv E_\alpha(\mathbf{R}_i) = \alpha_0^{-1} d_{i,\alpha}$, where $d_{i,\alpha}$ are defined in (3.33)–(3.35), $\mathbf{R}_i \equiv (x_i, y_i)$, and x_i and y_i are coordinates of the dipoles on the surface of a film). The results are shown for different values of frequency parameter X that using formula (3.30) can be related to wavelength, for any particular material. Note that local field

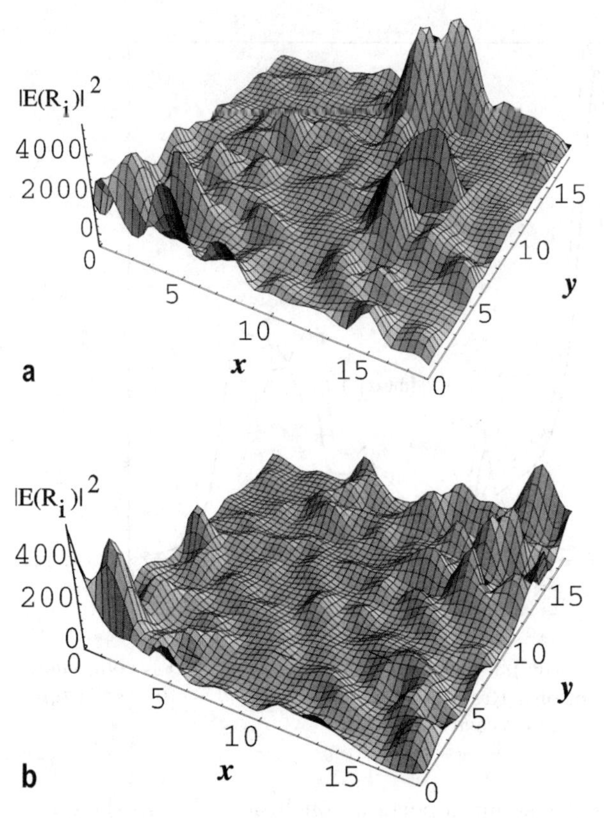

Fig. 4.3. Spatial distributions of the local field intensities $|\mathbf{E}(\mathbf{R}_i)|^2$ on a silver self-affine surface for different values of frequency parameter X: (**a**) $X = -3$ ($\lambda \approx 500\,\text{nm}$); (**b**) $X = -2$ ($\lambda \approx 400\,\text{nm}$). The decay parameter $\delta = 0.03$ in both cases. The applied field is polarized in the (x, y) plane, with $\mathbf{E}^{(0)} = (2)^{-1/2}(1, 1, 0)$

distributions $|\mathbf{E}(\mathbf{R}_i)|^2$ can be measured with the use of a near-field scanning optical microscope (NSOM), provided the probe is passive [124].

As seen in Fig. 4.3, for a modest value of $\delta = 0.03$ that is typical for metals in the visible and near-infrared parts of the spectrum, the local-field intensities in the hot zones can significantly - up to three orders of magnitude - exceed the intensity of the applied field (for smaller values of δ, enhancements can be even larger). The high frequency and polarization sensitivity of the field distributions is also obvious from the figure.

Strongly inhomogeneous distributions of local fields on a self-affine surface bring about large spatial fluctuations of local fields and strong enhancements of optical processes. As discussed in the next section, these enhancements are especially large for nonlinear optical phenomena, which are proportional to the local fields raised to a power greater than one.

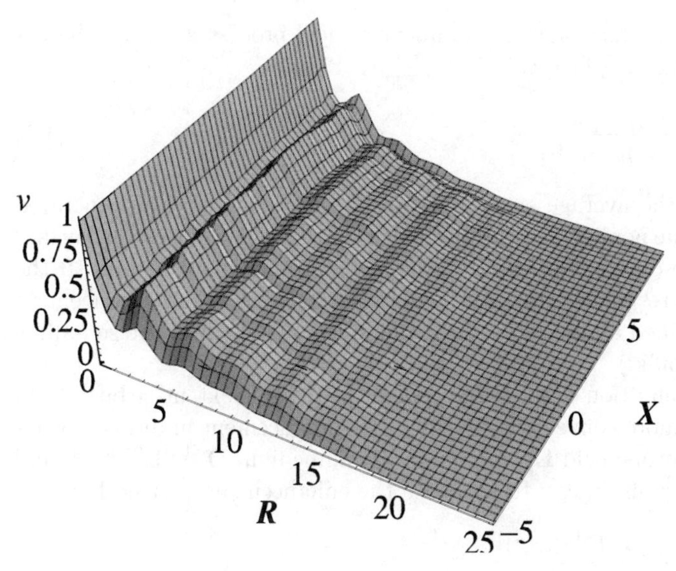

Fig. 4.4. The mode correlation function, $\nu(R, X)$

To study the localization of eigenmodes on a self-affine surface, we calculate the mode pair-correlation function defined as

$$\nu(R, X) = \left\langle C \sum_{i,j \in s; \alpha,\beta} \delta(R_{ij} - R)[(i\alpha|n)]^2[(j\beta|n)]^2 \right\rangle , \tag{4.3}$$

where the normalization constant C is defined by the requirement $\nu(R = 0) = 1$, and the summations are over dipoles on the surface only (averaging over a small interval ΔX near X is implied in the above formula). If individual (not averaged) mode n is localized within a certain area of radius R_0, then $\nu_n(R) = \nu(R, X = w_n)$ is small for $R > R_0$; the rate of decay of $\nu_n(R)$ at $R > R_0$ reflects a character of localization (strong or weak) for the state n [125].

The calculated $\nu(R, X)$ (see Fig. 4.4) is well approximated by the formula $\nu(R, X) = \exp\{-[R/L(X)]^\kappa\}$, where $\kappa \approx 0.7$. When the exponent κ is larger than unity, $\kappa > 1$, the modes are commonly called superlocalized; in our case, with $\kappa \approx 0.7$, the modes can be referred to as sublocalized (or quasilocalized), on average.

4.3 Enhanced Optical Phenomena on Self-Affine Surfaces

Now we consider enhancement of nonlinear optical processes on a self-affine surface, by using the quasi-static approximation. An average enhancement of the generated signal for coherent nonlinear optical processes can be characterized by the following factor:

$$G = \frac{|\langle \mathbf{D}^{\mathrm{NL}}(\omega_{\mathrm{g}})\rangle|^2}{|\mathbf{D}_0^{\mathrm{NL}}(\omega_{\mathrm{g}})|^2}, \tag{4.4}$$

where $\langle \mathbf{D}^{\mathrm{NL}}\rangle$ is the average surface-enhanced dipole moment at generated frequency ω_{g} of the nonlinear molecules when they are placed on a film surface and $\langle \mathbf{D}_0^{\mathrm{NL}}\rangle$ is the dipole moment of the same molecules in a vacuum. (If the nonlinear optical response is due to the material itself, then the enhancement factor can be defined as the ratio of the nonlinear signals from the self-affine surface and the bulk.)

Note that a definition of the enhancement is, to some extent, arbitrary. In some cases, it is more convenient to define the enhancement in terms of work done by a (real) probe field $\mathbf{E}^{(0)}(\omega_{\mathrm{g}})$ on a self-affine film, $\mathbf{D}^{\mathrm{NL}} \cdot \mathbf{E}^{(0)}(\omega_{\mathrm{g}})$, and in a vacuum, $\mathbf{D}_0^{\mathrm{NL}} \cdot \mathbf{E}^{(0)}(\omega_{\mathrm{g}})$. In this case the enhancement is given by

$$G = \frac{|\langle \mathbf{D}^{\mathrm{NL}}(\omega_{\mathrm{g}}) \cdot \mathbf{E}^{(0)}(\omega_{\mathrm{g}})\rangle|^2}{|\langle \mathbf{D}_0^{\mathrm{NL}}(\omega_{\mathrm{g}}) \cdot \mathbf{E}^{(0)}(\omega_{\mathrm{g}})\rangle|^2}. \tag{4.5}$$

The uniform probe field $\mathbf{E}_0(\omega_{\mathrm{g}})$ is assumed to be linearly polarized (so that it can be chosen real) and it should not be confused with the generated field $\mathbf{E}_i(\omega_{\mathrm{g}})$; the former produces the local field at ω_{g} through the linear relation (3.33). Note that enhancement in (4.5), does not depend on the magnitude of the probe field $\mathbf{E}^{(0)}(\omega_{\mathrm{g}})$. In many cases, a convenient choice for the polarization of the probe field $\mathbf{E}^{(0)}(\omega_{\mathrm{g}})$ is the same as the polarization of the nonlinear average dipole $\mathbf{D}^{\mathrm{NL}}(\omega_{\mathrm{g}})$. Formula (4.5) allows us to express enhancement factors in terms of the local fields only, as in Chap. 2.

4.3.1 Raman Scattering

Although spontaneous Raman scattering is a linear optical process, its enhancement on a surface is proportional to the local field raised to the fourth power and therefore, in terms of enhancement, it is similar to nonlinear processes. To obtain enhancement of Raman scattering from molecules when they are placed on a self-affine surface, we can follow the derivation presented in Chap. 3 for the case of fractal aggregates (see (3.50)–(3.56). This time, however, we take into account the difference between the applied and Stokes field frequencies (ω and ω_s respectively) and polarizabilities at these frequencies. Then, formula (3.55) can be rewritten as follows:

$$G_{\mathrm{RS}} = \frac{|Z_{\mathrm{s}} Z|^2}{|E^{(0)}|^2} \langle \alpha_{j,\beta\alpha}^{s} [\alpha_{j,\beta\gamma} E_{\gamma}^{(0)}] \alpha_{j,\delta\alpha}^{s*} [\alpha_{j,\delta\delta'}^{*} E_{\delta'}^{(0)*}] \rangle_{\mathrm{S}}, \tag{4.6}$$

where $Z_{\mathrm{s}} = Z(\omega_{\mathrm{s}}) = X_{\mathrm{s}} + \mathrm{i}\delta_{\mathrm{s}}$, and $\alpha_i^s \equiv \alpha_i(\omega_s)$, and summation over repeated indices is implied. The local polarizabilities are defined through the relation $d_{i,\alpha} = \alpha_{i,\alpha\beta} E_{\beta}^{(0)}$ and can be found by solving the CDE. The terms in the square brackets in (4.6) are proportional to the local fields at the fundamental frequency ω acting on the jth site. The sign $\langle ... \rangle_S$ in (4.6) denotes an average over positions of dipoles on the surface of the film only. Note that the enhancement can be also roughly estimated as

$$G_{RS} \sim \left\langle \frac{|E_i(\omega)E_i(\omega_s)|^2}{|E^{(0)}|^4} \right\rangle_S . \tag{4.7}$$

Figure 4.5 shows the average enhancement of Raman scattering for small and large Stokes shifts (see Sect. 3.6.2) on self-affine silver films generated in the RSS model. $G_{RS,\parallel}$ and $G_{RS,\perp}$ describe enhancements for the applied field, polarized in the plane of the film and perpendicular to it respectively. For the dielectric function of silver $\epsilon(\lambda)$ the data of [89] was used. Results of the calculations were averaged over 12 random samples with $N \sim 10^3$ dipoles in each sample. As seen in the figure, enhancement increases toward the long-wavelength part of the spectrum and reaches very large values, $\sim 10^7$; this agrees well with experimental observations of SERS on rough thin films [51].

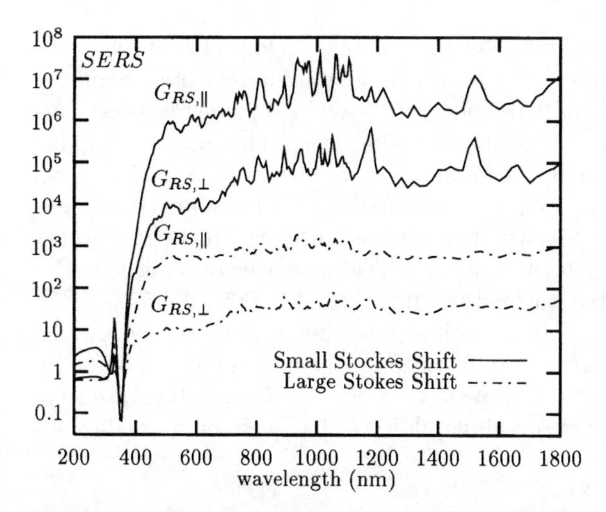

Fig. 4.5. Enhancement factor for Raman scattering, $G_{RS,\parallel} = [G_{RS,x} + G_{RS,y}]/2$ and $G_{RS,\perp} = G_{RS,z}$, on silver self-affine films for small and large Stokes shifts

In Fig. 4.6, the field spatial distributions at the fundamental and Stokes frequencies are shown [115] (where we set $|Z_s\zeta|^2 = 0.25$, with *zeta* being the Raman polarizability). The local intensity of the Stokes field $I_i^s = I^s(\mathbf{r}_i) = |\mathbf{E}_i^s|^2$ was found using the formula [115] $I_i^s = |Z_s|^4|\zeta|^2 \sum_{j\in S} |\alpha_{ij,\alpha\beta}^s E_{j,\beta}|^2$, where the summation is over dipoles on the surface only, and $\alpha_{ij,\alpha\beta}^s$ is defined through the relation $\alpha_{i,\alpha\beta}^s = \sum_j \alpha_{ij,\alpha\beta}^s$, with $\alpha_{i,\alpha\beta}^s$ being the linear polarizability at Stokes frequency (see (3.34)).

As seen in Fig. 4.6, the distributions are found to contain hot spots, where the fields are very high. The spatial positions of these spots are strong func-

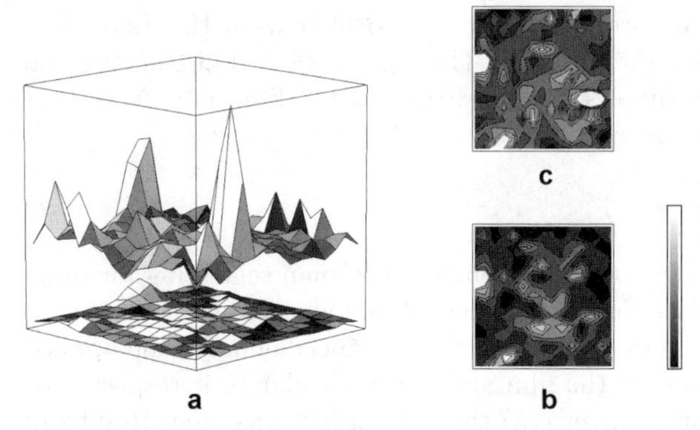

Fig. 4.6. (**a**) Spatial distributions of the local fields at the fundamental frequency $\lambda = 550\,\mathrm{nm}$ (*bottom*; the field distribution being magnified by a factor of 3) and for the Stokes fields $\lambda_s = 600\,\mathrm{nm}$ (*top*). The applied field is linearly polarized in the plane of the film. (**b**) and (**c**) The contour-plots for the field distributions shown on (**a**) (*bottom*) and (**a**) (*top*) respectively

tions of the frequency and polarization of the applied field [114]. Although the Stokes signal is proportional to the local field at the fundamental frequency ω, the generated Stokes field at frequency ω_s, excites, in general, other surface eigenmodes. Hence the field spatial distributions produced by the applied field and by the Raman signal can be different, as clearly seen in the figure. (In the field distributions at frequencies ω and ω_s, the hot spots at the left side of Figs. 4.6b and 4.6c are spatially correlated, whereas the largest Stokes peak is positioned at the right, where the ω-field is relatively small.)

This picture is expected to be typical for various optical processes in strongly disordered fractal systems, such as self-affine thin films. Specifically, hot spots associated with fields at different frequencies and polarizations can be localized in spatially separated nm-sized areas. The NSOM observations of optical excitations on a silver self-affine film [47, 77] verify basic predictions for the field distribution.

4.3.2 Second-Harmonic Generation

In this section, we consider second-harmonic generation (SHG) from self-affine surfaces as a typical example of nonlinear optical phenomena of the second order [126]. Since, in Chap. 3, high-order harmonic generation was only briefly mentioned, here we discuss second- and third-harmonic generation in more detail.

Because SHG is extremely sensitive to surface roughness conditions, it has become a widely used technique for studying structural and electronic properties of surfaces and interfaces [126–128]. Thus this process is of fundamental importance for understanding the nonlinear interaction of light with thin films.

For SHG, nonlinear polarization $\mathbf{P}^{(2)}(2\omega)$ is commonly introduced through the nonlinear susceptibility tensor of a third rank $\hat{\boldsymbol{\chi}}^{(2)}(-2\omega;\omega,\omega)$ as

$$\mathbf{P}^{(2)}(2\omega) = \hat{\boldsymbol{\chi}}^{(2)}(-2\omega;\omega,\omega) : \mathbf{E}(\omega) : \mathbf{E}(\omega). \tag{4.8}$$

(Note that we use here slightly different notations for a tensor product.)

Extensive studies of different mechanisms of SHG on surfaces [129, 130] and in bulk [131, 132] have been carried out for jellium and other models since they have been proposed by Rudnick and Stern [133]. Here, for simplicity, we assume that contributions to $\hat{\boldsymbol{\chi}}^{(2)}$ are associated with adsorbed molecules placed on the film's surface. The nonlocal effects related to the spatial dispersion and the effects of a finite depth field penetration are left out of the analysis.

We adopt the following asymmetrical structure for the non-linear molecules on the surface. They are assumed to have a "preferred" direction $\mathbf{n}$, which coincides with the vector normal to the (x, y) plane of the film, so that the film anisotropy is reproduced by the adsorbed molecules [134].

We construct the vector $\mathbf{P}^{(2)}(2\omega)$ from the obvious independent combinations of the triplet $(\mathbf{n}, \mathbf{E}, \mathbf{E})$ in (4.8) that involve the field squared:

$$\mathbf{P}^{(2)}(2\omega) = A(\mathbf{E}\cdot\mathbf{E})\mathbf{n} + B(\mathbf{n}\cdot\mathbf{E})\mathbf{E} + C(\mathbf{n}\cdot\mathbf{E})^2\mathbf{n}, \tag{4.9}$$

where A, B, and C are three independent complex constants determined only by the internal structure of the molecules on the surface.

Comparing the components of the polarization vector given in (4.8) with the ones introduced in (4.9), we obtain the following relations for the non-zero components of $\hat{\boldsymbol{\chi}}^{(2)}$:

$$\begin{aligned}
&\chi^{(2)}_{xxz} = \chi^{(2)}_{xzx} = \chi^{(2)}_{yzy} = \chi^{(2)}_{yyz} = B/2,\\
&\chi^{(2)}_{zxx} = \chi^{(2)}_{zyy} = A\\
&\chi^{(2)}_{zzz} = A + B + C.
\end{aligned} \tag{4.10}$$

The amplitude of a nonlinear dipole located at the ith site can be written similarly to (4.9):

$$\mathbf{d}_i^{\mathrm{NL}} = a(\mathbf{E}_i\cdot\mathbf{E}_i)\mathbf{n} + b(\mathbf{n}\cdot\mathbf{E}_i)\mathbf{E}_i + c(\mathbf{n}\cdot\mathbf{E}_i)^2\mathbf{n} \tag{4.11}$$

where $a = Av$, $b = Bv$, $v = 4\pi R_{\mathrm{m}}^3/3 = a^3$, and $\mathbf{E}_i$ is the local field.

First, we find the average nonlinear dipole moment $\langle\mathbf{D}_0^{\mathrm{NL}}\rangle$ for the molecules responsible for SHG on the plane with $\epsilon \approx 1$ (in a vacuum). In this case, the induced dipoles are excited only by uniform incident field $\mathbf{E}_0$, so that $\mathbf{E}_i = \mathbf{E}^{(0)}$. It follows from (4.11) that

$$\langle\mathbf{D}_0^{\mathrm{NL}}\rangle = a(\mathbf{E}^{(0)}\cdot\mathbf{E}^{(0)})\mathbf{n} + b(\mathbf{n}\cdot\mathbf{E}^{(0)})\mathbf{E}^{(0)} + c(\mathbf{n}\cdot\mathbf{E}^{(0)})^2\mathbf{n}. \tag{4.12}$$

Vector $\mathbf{P}^{(2)}(2\omega)$ depends on the light polarization; here we assume that an incident wave is linearly polarized. For the incident wave that is polarized along the z-axis (hereafter we refer to this as p-polarization; note that z-polarization is a special case of p-polarization) or in the (x, y) plane (s-polarization), we obtain

$$|\langle \mathbf{D}_0^{\mathrm{NL}} \rangle|^2 = |\mathbf{E}^{(0)}|^4 |a+b+c|^2, \text{ for } p\text{-polarization, and,} \tag{4.13}$$

$$|\langle \mathbf{D}_0^{\mathrm{NL}} \rangle|^2 = |\mathbf{E}^{(0)}|^4 |a|^2, \text{ for } s\text{-polarization.} \tag{4.14}$$

Because the induced nonlinear dipoles on the metal surface interact with each other via the *linear* polarizabilities $\alpha_0(2\omega)$, we can write an analog of the CDE (3.25) for the nonlinear dipole amplitudes as

$$\mathbf{d}_i^{\mathrm{NL}} = v\hat{\boldsymbol{\chi}}^{(2)}(2\omega;\omega,\omega) : \mathbf{E}_i^2(\omega) + \alpha_0(2\omega) \sum_{j \neq i} \hat{W}_{ij} \mathbf{d}_j^{\mathrm{NL}}(2\omega) \tag{4.15}$$

and its matrix counterpart

$$|d^{\mathrm{NL}}) = v|\hat{\chi}^{(2)} : E^2) + \alpha_0(2\omega)\hat{W}|d^{\mathrm{NL}}), \tag{4.16}$$

where

$$v(i\alpha|\hat{\chi}^{(2)} : E^2) = an_\alpha E_{i\beta}E_{i\beta} + bn_\beta E_{i\alpha}E_{i\beta} + cn_\alpha n_\beta n_\gamma E_{i\beta}E_{i\gamma}. \tag{4.17}$$

In (4.15) and (4.16) we include the linear interaction of the local nonlinear dipoles at the double frequency 2ω. This provides an additional contribution to the nonlinear SHG signal.

The interactions of nonlinear dipoles at generated frequency 2ω become important if this frequency is within the surface mode band. In this case, the local fields at both the fundamental and generated frequencies can excite the resonance surface modes and thus get strongly enhanced. Accordingly, the resultant enhancement can become much larger than for the case without interaction of nonlinear dipoles. In order to take into account the coupling of nonlinear dipoles, we must use equation (4.15). A formal solution to a similar equation for linear dipole moments (3.25) is given in Chap. 3 (see (3.35)). It can be easily generalized for the case of the coupled nonlinear dipoles (4.16); then in matrix form the solution is

$$|d^{\mathrm{NL}}) = vZ(2\omega) \sum_n \frac{|n)(n|\hat{\chi}^{(2)} : E^2)}{Z(2\omega) - w_n}, \tag{4.18}$$

where $Z(2\omega) \equiv \alpha_0^{-1}(2\omega)$ and $\alpha_0(2\omega)$, as mentioned, is the scalar polarizability of a monomer at the double frequency.

Using (4.17), (4.18), and (3.33), we express the Cartesian components of the average nonlinear dipole moment $\langle \mathbf{D}^{\mathrm{NL}}(2\omega) \rangle$ as

$$\begin{aligned} \langle D_\alpha^{\mathrm{NL}} \rangle = Z(2\omega)Z^2(\omega) \Big[&(a+b+c)\langle \alpha_{z\alpha} d_z^2 \rangle + a\langle \alpha_{z\alpha}(d_x^2 + d_y^2) \rangle \\ &+ b\langle \alpha_{x\alpha} d_x d_z + \alpha_{y\alpha} d_y d_z \rangle \Big], \end{aligned} \tag{4.19}$$

where $\alpha_{\alpha\beta} \equiv \alpha_{i,\alpha\beta}(\omega_{\mathrm{g}}) \equiv \alpha_{\alpha\beta}(\omega_{\mathrm{g}})$ is the local linear polarizability at the generated frequency ω_{g}. (For simplicity, we omit the subscript i and argument 2ω in $\alpha_{i,\alpha\beta}(\omega_{\mathrm{g}})$ in (4.19).) The linear polarizability at the generated frequency is defined similarly to (3.34) as

$$\alpha_{i,\alpha\beta}(\omega_{\mathrm{g}}) = \sum_{j,n} \frac{(i\alpha|n)(n|j\beta)}{Z(\omega_{\mathrm{g}}) - w_n}, \quad i \in S. \tag{4.20}$$

The summation over j, i.e. over all the monomers, in (4.20) is the consequence of the coupling of nonlinear dipoles given by the second term in (4.15).

The average (over a film surface) SHG enhancement factor can be obtained by substituting (4.13), (4.14) and (4.19) in (4.4), which gives:

$$G_{\mathrm{SHG}} = \frac{|Z(2\omega)|^2}{|E^{(0)}|^4} \sum_{\beta} \Big| \langle \alpha_{z\beta} E_z^2 \rangle + \frac{1}{\Gamma} \langle \alpha_{z\beta}(E_x^2 + E_y^2) \rangle$$
$$+ \left(1 - \frac{1}{\Gamma}\right) \langle \alpha_{x\beta} E_x E_z + \alpha_{y\beta} E_y E_z \rangle \Big|^2, \text{ for } p\text{-polarization, and}$$
$$G_{\mathrm{SHG}} = \frac{|Z(2\omega)|^2}{|E^{(0)}|^4} \sum_{\beta} \Big| \Gamma \langle \alpha_{z\beta} E_z^2 \rangle + \langle \alpha_{z\beta}(E_x^2 + E_y^2) \rangle$$
$$+ (\Gamma - 1) \langle \alpha_{x\beta} E_x E_z + \alpha_{y\beta} E_y E_z \rangle \Big|^2, \text{ for } s\text{-polarization,} \tag{4.21}$$

where $\alpha_{\alpha\beta} = \alpha_{i,\alpha\beta}(\omega_{\mathrm{g}}) = \alpha_{\alpha\beta}(2\omega)$ is given by (4.20), E_α represents the local field components, and the "oblique coefficient" Γ is defined with the use of (4.10) as

$$\Gamma \equiv \frac{\chi^{(2)}_{zzz}}{\chi^{(2)}_{zxx}} = \frac{\chi^{(2)}_{zzz}}{\chi^{(2)}_{zyy}} = 1 + (b + c)/a. \tag{4.22}$$

Note that the above expressions contain the nonlinear polarizability tensor $\hat{\alpha}_i(2\omega)$ and cannot be written only in terms of the local fields $\mathbf{E}_i$. Therefore, if interactions of the generated nonlinear dipoles at frequency ω_{g} are important, it is impossible in general to express the enhancement factor in terms of the local and incident fields only.

If we use definition (4.5) to express the enhancement in terms of work done by a probe field at the generated frequency, then for the case of p-polarization, for example, we obtain the following expression:

$$G_{\mathrm{SHG}} = \frac{1}{|E^{(0)2} E^{(0)}(2\omega)|^2} \Big| \langle E_z(2\omega) E_z^2 \rangle + \frac{1}{\Gamma} \langle E_z(2\omega)(E_x^2 + E_y^2) \rangle$$
$$+ \left(1 - \frac{1}{\Gamma}\right) \langle E_x(2\omega) E_x E_z + E_y(2\omega) E_y E_z \rangle \Big|^2, \tag{4.23}$$

where E is the local field at frequency ω, and $E(2\omega)$ is the local *linear* field at 2ω, which is related to the probe field $E^{(0)}(2\omega)$ as $E_{i,\alpha}(2\omega) =$

$Z(2\omega)\alpha_{i,\alpha\beta}(2\omega)E_{\beta}^{(0)}(2\omega)$, with $\alpha_{i,\alpha\beta}(2\omega)$ defined in (4.20). Although the enhancement does not depend on the magnitude of $E^{(0)}(2\omega)$, it depends in general on the polarization for $\mathbf{E}^{(0)}(2\omega)$. Note that the above formulas for SHG enhancement are consistent with the estimate formula (2.4) but they take into account the geometry of interaction.

Equations (4.21) and (4.23) can be simplified when the coupling of nonlinear dipoles is not effective on the surface. As mentioned, that occurs when the generated frequency $\omega_{\mathrm{g}} = 2\omega$ is far from any of the surface eigenmodes, so that $|Z(\omega_{\mathrm{g}})| \gg w_n, \ \forall n$. Then the polarizability matrix (4.20) becomes diagonal: $\alpha_{i,\alpha\beta}(\omega_{\mathrm{g}}) \approx \delta_{\alpha\beta}/Z(\omega_{\mathrm{g}})$. In this case, we can set $E(2\omega) = E^{(0)}(2\omega)$ in (4.23)).

We also note that contributions from the last terms in (4.21) and (4.23), which are proportional to strongly fluctuating (and sign-changing) products E_xE_z and E_yE_z, are small and can be neglected with good accuracy. Also, in the short-wavelength part of the spectrum, when the interaction of the dipoles at 2ω can, typically, be neglected (because the surface modes are not excited), the diagonal element $\alpha_{zz}(2\omega)$ in (4.21) and (4.23) dominates the off-diagonal ones, so that the local dipoles at 2ω are directed along the corresponding local fields at 2ω. In contrast, in the long-wavelength part of the spectrum, when the oscillations at 2ω lie within the surface mode band, the largest dipoles at 2ω are excited when the light is polarized in the plane of the film, and therefore the off-diagonal elements $\alpha_{zx}(2\omega)$ and $\alpha_{zy}(2\omega)$ are larger than $\alpha_{zz}(2\omega)$. These arguments allow simplification of the general expressions (4.21) and (4.23) when they are used in the corresponding limiting cases.

Figure 4.7 shows the averaged (over 12 different film samples) enhancement factor for SHG from silver self-affine surfaces; the enhancement was calculated using formulas (4.21). Results are shown for $\Gamma = 2$. Note that real values of Γ indicate an absence of nonlinear absorption by molecules. In [116], it was found that, for different numerical values of Γ, the enhancement factors G are similar in terms of their magnitudes and spectral dependences. This is in agreement with the fact that the spectral dependence of linear dipoles in (3.35) does not correlate with the value of Γ chosen to be independent of the incident wavelength.

We see that the anticipated inequality $G_{\|} \gg G_{\perp}$ holds, since the linear dipoles and corresponding local fields (3.36) are, on average, larger for the incident field polarized in the plane of the film than in the normal direction; this is because a thin film can be roughly thought of as an oblate spheroid with a high aspect ratio. The largest average enhancement for SHG is $\sim 10^7$.

In [114, 116] it was also found that, in the quasi-static limit, both linear and nonlinear properties of self-affine films do not depend on the linear size of the system and the numbers of monomers.

Below we also consider spatial distributions of local enhancements on a film surface, $g_{\mathrm{SHG}}(\mathbf{r}_i) = |\mathbf{d}_i^{\mathrm{NL}}|^2 \big/ |\mathbf{d}_0^{\mathrm{NL}}|^2, \ \ i \in S$. The nonlinear local dipoles

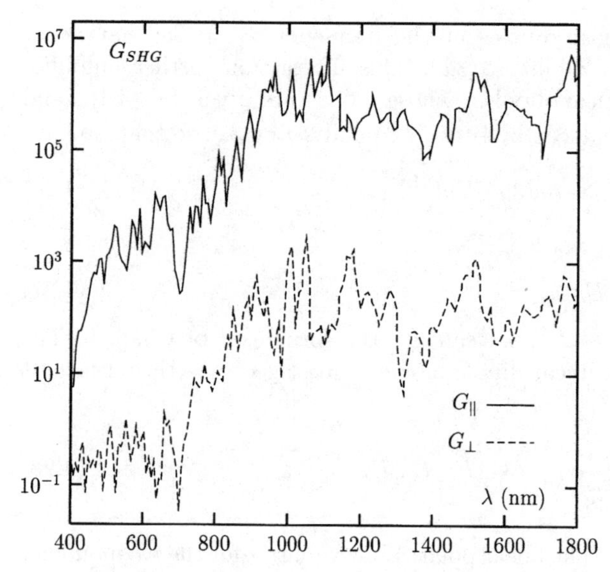

Fig. 4.7. Average enhancement factors for second-harmonic generation (SHG) from a self-affine silver surface, for light polarized in the (x, y)plane of the film ($G_{\mathrm{SHG}} \equiv G_{\parallel}$) and in the normal z-direction ($G_{\mathrm{SHG}} \equiv G_{\perp}$)

on a metal self-affine surface, $\mathbf{d}_i^{\mathrm{NL}} \equiv \mathbf{d}_i^{\mathrm{NL}}(2\omega)$, and those on a surface with $\epsilon = 1$ (i.e., in a vacuum), $\mathbf{d}_0^{\mathrm{NL}} \equiv \mathbf{d}_0^{\mathrm{NL}}(2\omega)$, are given by (4.18) with (4.17) and (4.12) respectively. The spatial distribution of SHG will be compared below with that for third-harmonic generation.

4.3.3 Third-Harmonic Generation

We now consider third-order optical susceptibilities, assuming that they are associated with spherically symmetrical molecules adsorbed on a self-affine surface. Such symmetry implies, in particular, that there is only one independent component for the fourth-rank susceptibility tensor $\chi^{(3)}_{\alpha\beta\gamma\delta}(-3\omega; \omega, \omega, \omega)$, which is responsible for third-harmonic generation (THG).

The spherical symmetry of molecules on the surface allows us to express the amplitudes of the nonlinear dipole moments as

$$\mathbf{d}_i^{\mathrm{NL}}(3\omega) = c\mathbf{E}_i(\mathbf{E}_i \cdot \mathbf{E}_i), \tag{4.24}$$

where c is the only independent element of the third-order susceptibility tensor. In (4.24), we neglect interactions of the nonlinear dipoles at the frequency 3ω. Using (4.24) and replacing $\mathbf{E}_i$ by $\mathbf{E}^{(0)}$, we find the denominator in (4.4) expressed as

$$|\langle \mathbf{D}_0^{\mathrm{NL}} \rangle|^2 = |c|^2 |\mathbf{E}^{(0)}|^6. \tag{4.25}$$

Interaction of nonlinear dipoles at the frequency $\omega_g = 3\omega$ may occur through the linear polarizability $\alpha(3\omega)$. This interaction further amplifies amplitudes of the nonlinear dipoles, whose values are given by (4.16) and (4.17) with $|\hat{\chi}^{(2)} : E^2)$ replaced by $|\hat{\chi}^{(3)} : E^3)$ and 2ω by 3ω, so that we have

$$|d^{\rm NL}) = v|\hat{\chi}^{(3)} : E^3) + \alpha_0(3\omega)\hat{W}|d^{\rm NL}), \tag{4.26}$$

where

$$v(i\alpha|\hat{\chi}^{(3)} : E^3) = cE_{i\alpha}E_{i\beta}E_{i\beta}. \tag{4.27}$$

To solve (4.26) and (4.27), we can use the formalism of Chap. 3. The α–component of the nonlinear dipole moment, averaged over the surface, is given by

$$\langle D_\alpha^{\rm NL}\rangle = \frac{cZ(3\omega)}{N_{\rm S}}\sum_{j\in S}\alpha_{j,\beta\alpha}(3\omega)E_{j\beta}E_{j\gamma}E_{j\gamma}, \tag{4.28}$$

where $Z(3\omega) = \alpha_0^{-1}(3\omega)$. The linear polarizability matrix for the ith monomer is defined by (4.20).

It is interesting to note that $\langle D_\alpha^{\rm NL}\rangle$ is in general complex, even if c is real, i.e. even if there is no nonlinear absorption in the system. This property reflects the fact that the average nonlinear dipole moment $\langle \mathbf{D}^{\rm NL}\rangle$ is affected by the surface eigenmodes having finite losses. Also, $\langle \mathbf{D}^{\rm NL}\rangle$ is elliptically polarized because the complex matrix (3.34) transforms the linear polarization of the incident field into an elliptical polarization of dipole moments (see also formula (3.33)).

Substitution of (4.25) and (4.28) in (4.4) gives the following expression for the surface-enhanced THG in the case of linear polarization of the incident wave:

$$G_{\rm THG} = \frac{|Z(3\omega)|^2}{|E^{(0)}|^6}\sum_\gamma |\langle \alpha_{\beta\gamma}(3\omega)E_\beta(\mathbf{E}\cdot\mathbf{E})\rangle|^2. \tag{4.29}$$

If we use formula (4.5) for the enhancement factor, the latter can be expressed in terms of the local fields as

$$G_{\rm THG} = \frac{\left|\left\langle \Big(\mathbf{E}(3\omega)\cdot\mathbf{E}\Big)\Big(\mathbf{E}\cdot\mathbf{E}\Big)\right\rangle\right|^2}{|E^{(0)3}E^{(0)}(3\omega)|^2}, \tag{4.30}$$

where E and $E(3\omega)$ are the local linear fields at frequencies ω and 3ω, induced by the applied field $E^{(0)}$ and the probe field $E^{(0)}(3\omega)$ respectively (cf. (2.5)).

If the generated signal 3ω does not excite the surface eigenmodes, so that $\alpha_{\alpha\beta}(3\omega) \approx \delta_{\alpha\beta}/Z(3\omega)$, expression (4.29) simplifies to

$$G_{\rm THG} = \frac{|\langle \mathbf{E}(\mathbf{E}\cdot\mathbf{E})\rangle|^2}{|E^{(0)}|^6}. \tag{4.31}$$

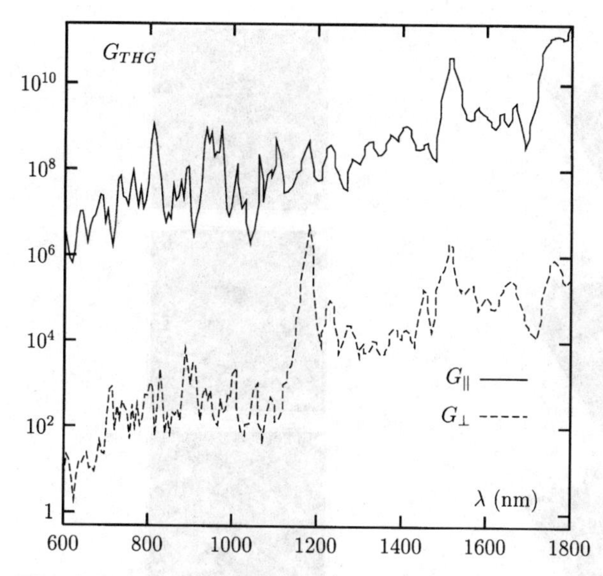

Fig. 4.8. Average enhancement factors for third-harmonic generation (THG) from a self-affine silver surface, for light polarized in the (x, y)plane of the film ($G_{\text{THG}} \equiv G_{\|}$) and in the normal z-direction ($G_{\text{THG}} \equiv G_{\perp}$)

Figure 4.8 shows enhancement factor G_{THG} calculated using formula (4.29). The values of G_{THG} are even larger than for G_{SHG}, reaching $\sim 10^{11}$ values. THG involves the fields to the higher powers, so that the dominance of local fields $\mathbf{E}_i$ over $\mathbf{E}^{(0)}$ leads to the larger values of enhancement.

We can also calculate spatial distributions of local enhancements of THG on a film surface defined as $g_{\text{THG}}(\mathbf{r}_i) = |\mathbf{d}_i^{\text{NL}}|^2/|\mathbf{d}_0^{\text{NL}}|^2, \quad i \in S$, where $\mathbf{d}_i^{\text{NL}}$ and $\mathbf{d}_0^{\text{NL}}$ are the local nonlinear dipoles at 3ω on metal self-affine surface and in a vacuum respectively.

In Figs. 4.9a,b, spatial distributions of local-field enhancements at the fundamental frequency, $g = |E/E^{(0)}|^2$ and local enhancements of SHG and THG, $g_{\text{SHG}} = |\mathbf{d}_i^{\text{NL}}(2\omega)|^2/|\mathbf{d}_0^{\text{NL}}(2\omega)|^2$ and $g_{\text{THG}} = |\mathbf{d}_i^{\text{NL}}(3\omega)|^2/|\mathbf{d}_0^{\text{NL}}(3\omega)|^2$, are shown. The distributions of local enhancements are calculated for two wavelengths, 1 μm (Fig. 4.9a) and 10 μm (Fig. 4.9b), for the light polarized in the plane of the film. As discussed above, largest average enhancements are achieved in the infrared spectral range for the s-polarized incident light.

In the counter-plots of Fig. 4.9, the white spots correspond to higher intensities whereas the dark areas represent the low-intensity zones. We can see that spatial positions of the hot and cold spots in the enhancement distributions at the fundamental and generated frequencies are localized in small spatially separated parts of the film. Since the fundamental and generated frequencies are different, the incident and generated waves excite different optical modes of the film surface and therefore produce different local field distributions. With the frequency alternation, the locations of the hot and

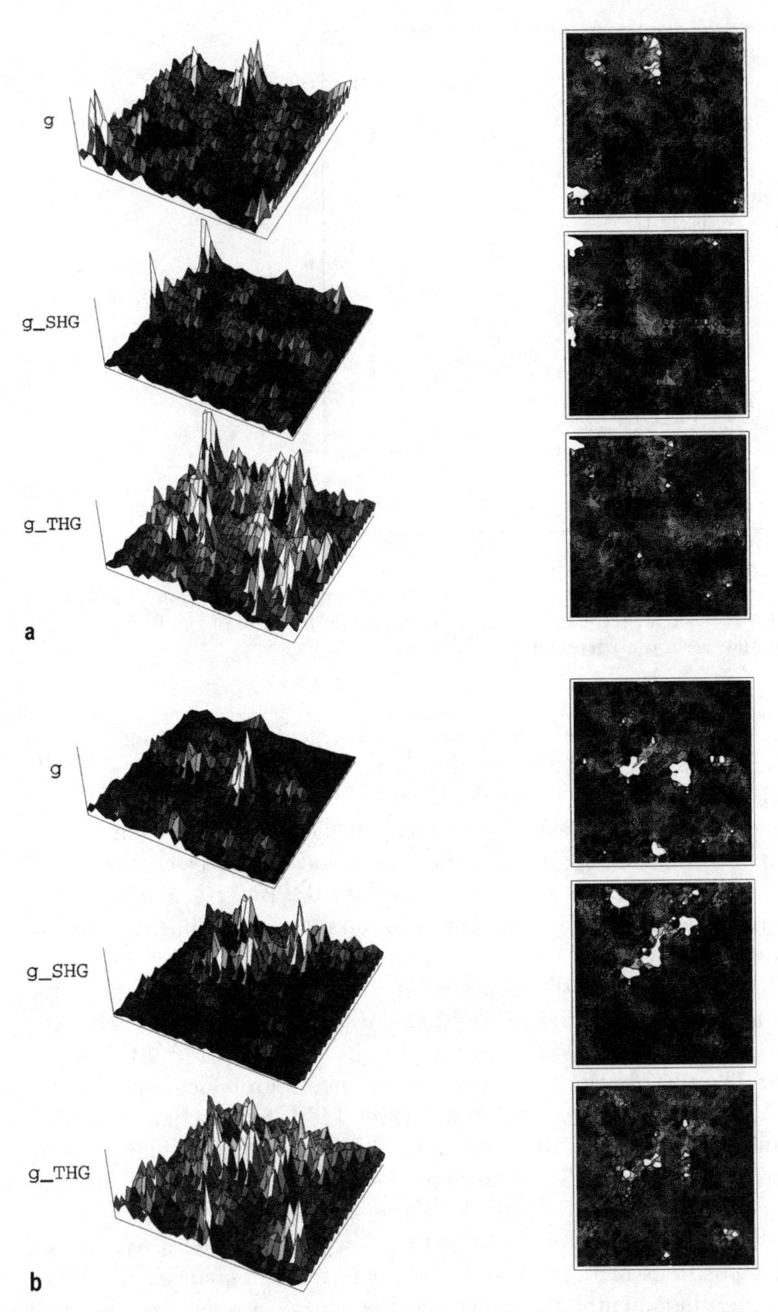

Fig. 4.9. Spatial distributions of the local enhancements for the field at the fundamental wavelength g for an SHG signal, g_{SHG}, and for a THG signal, g_{THG}. The corresponding counter-plots for the spatial distributions are also shown. (**a**) The fundamental wavelength is $\lambda = 1\,\mu\mathrm{m}$. The linear scales are used in all cases. The highest enhancement values in the figures are as follows: $g = 5\times10^3$, $g_{\mathrm{SHG}} = 5\times10^8$ and $g_{\mathrm{THG}} = 2\times10^{12}$. (**b**) Same as in (**a**) but for $\lambda = 10\,\mu\mathrm{m}$. The highest enhancement values are as follows: $g = 3\times10^4$, $g_{\mathrm{SHG}} = 10^{13}$ and $g_{\mathrm{THG}} = 2\times10^{19}$

cold spots change for all the fields at the fundamental and generated frequencies. Thus, different waves involved in the nonlinear interactions in a self-affine thin film produce nanometer-sized hot spots spatially separated for different waves. A similar effect was considered in Subsect. 4.3.1 for Raman scattering from self-affine films.

The magnitudes of the local field intensities in Fig. 4.9, grow with the wavelength. The highest local enhancement factor in the spatial distribution g changes from 5×10^3 at $\lambda = 1\,\mu m$ to 3×10^4 at $\lambda = 10\,\mu m$. For SHG and THG spatial distributions, the maximum enhancement factor increases respectively from 5×10^8 to 10^{13} and from 2×10^{12} to 2×10^{19}. Such behavior correlates with the fact that the average enhancement factor increases toward the infrared spectral region.

It is important to emphasize that local enhancements can exceed the average by several orders of magnitude. For example, comparison of the maximum local enhancement with the average enhancement for $\lambda = 1\,\mu m$ shows that the maximum intensity peaks exceed the average intensity by approximately two orders of magnitude for SHG (cf. Figs. 4.7 and 4.9) and by four orders of magnitude for THG (cf. Figs. 4.8 and 4.9). This occurs, in part, because the spatial separation between the hot spots can be significantly larger than their characteristic sizes; and it also occurs, in part, due to the destructive interference between the fields generated in different peaks. This destructive interference is typically large for the processes that do not include subtraction (annihilation) of photons of the incident waves in one elementary act of the nonlinear interaction of photons, such as the processes of the nth harmonic generation $\omega_s = n\omega$ or sum-frequency generation $\omega_s = \omega_1 + \omega_2 + ... + \omega_n$ [45] (see also Subsect. 3.6.1).

The giant local enhancements of nonlinear processes (e.g. up to 10^{19} for THG at $10\,\mu m$) open a fascinating possibility of fractal-surface-enhanced nonlinear optics and spectroscopy of single molecules. Also, if near-field scanning optical microscopy is employed, nonlinear nanooptics and nanospectroscopy with nanometer spatial resolution become possible. In contrast, conventional far-zone optics allows measurements of only an average enhancement of optical processes.

4.3.4 Kerr Optical Nonlinearity and Four-Wave Mixing

The enhancement factors for four-wave mixing (FWM) and for Kerr optical nonlinearity responsible for the nonlinear refraction and absorption in a medium are given by (2.2) and (2.1) respectively, and were considered in detail in Chap. 3. We use these formulas to calculate the enhancements. As above, the local fields (3.36) can be found by solving the coupled-dipole equations (3.25) for the system of dipoles obtained using the discrete dipole approximation and the RSS model of a self-affine surface.

In Fig. 4.10, the magnitude of the real part of the Kerr enhancement factor on silver self-affine films is shown as a function of the wavelength. (Note that

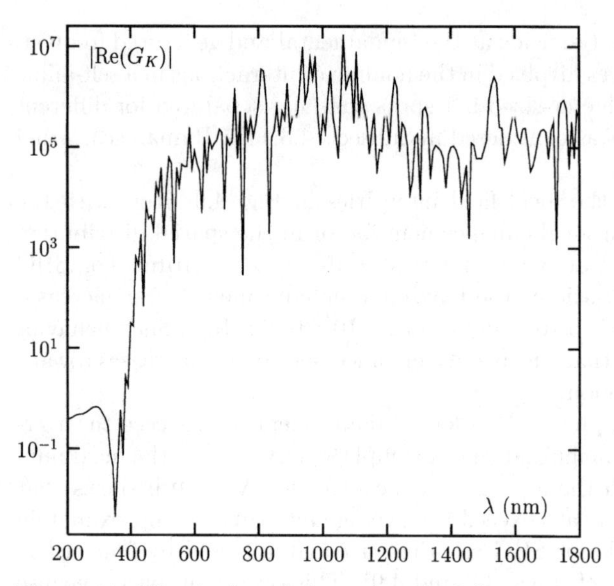

Fig. 4.10. Magnitudes of the real part of the average enhancement factor of the Kerr-nonlinearity $|G'_{\rm K}|$ for the light polarized in the (x, y) plane of a film

$G'_{\rm K}$ is negative for most of the frequencies and exceeds in magnitude the imaginary part $G''_{\rm K}$ [116]).

The huge average enhancement for degenerate FWM (DFWM) on a silver self-affine film is illustrated in Fig. 4.11, for the applied field polarized in the (x, y)-plane and perpendicular to it. The larger values of the average enhancement for DFWM, compared with THG, are explained by the fact that the interaction of nonlinear dipoles is stronger when the generated frequency is equal to the fundamental one. Also, the role of destructive interference for the field generated in different points is much larger for high-order harmonic generation than for DFWM, as discussed above.

The calculations show that both $G_{\rm K}$ and $G_{\rm FWM}$ are especially large toward the infrared part of the spectrum, where the local fields are larger and optical excitations are, on average, more localized.

We also mention here that strong enhancement of two-photon electron emission from the rough self-affine surface of a metal film has been predicted and obtained [116].

It is worth stressing that the giant enhancements of optical phenomena described above are related to the surface fractal geometry. As shown, in self-affine thin films, resonance optical excitations tend to be localized in small sub-wavelength regions, so that the local fields vary (fluctuate) strongly from hot to cold spots on the surface of the film. These large and strongly fluctuating local fields provide the giant enhancements of the nonlinear optical processes.

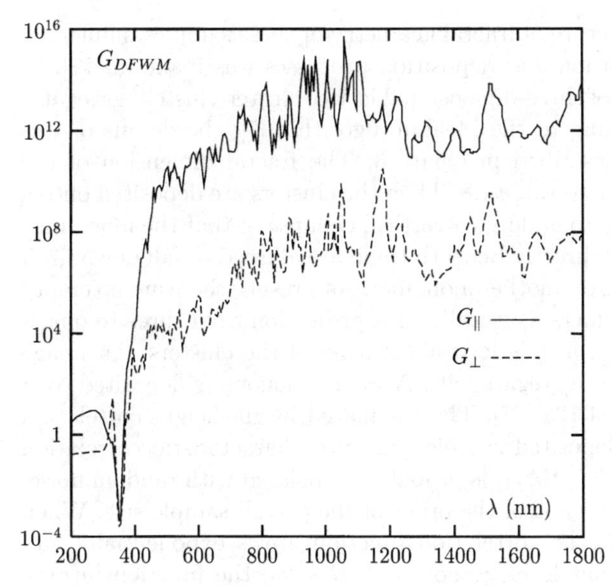

Fig. 4.11. Average DFWM enhancement factors from a self-affine silver surface, for light polarized in the (x, y) plane of a film ($G_{\mathrm{DFWM}} \equiv G_{\|}$) and in the normal z-direction ($G_{\mathrm{DFWM}} \equiv G_{\perp}$)

4.4 Nano-Optics of Fractal Aggregates Deposited on a Surface

4.4.1 Sample Preparation and Numerical Simulations

The self-affine surfaces considered above can be produced, for example, by atomic deposition onto a cold-temperature substrate; the restricted solid-on-solid model used here is a good numerical model for such films.

Another type of self-affine surface can be made by gravitationally depositing fractal cluster-cluster aggregates (CCAs) onto a substrate out of a solution. This idea was originally proposed by Moskovits et al. [47]. In this method, fractal aggregates are first prepared in solution as described in Chapter 3, and then gravitationally deposited onto (for example) a glass substrate, where the water is soaked out. It should be noted that the process of deposition of the samples on a surface and consequent evaporation of water results in a significant restructuring of the original self-similar fractal CCAs. Especially, the restructuring is important in the direction normal to the prism surface because of the collapse that can be caused by the capillary forces and a projection to two dimensions from three. However, as shown below, there are several indications that the samples obtained as a result of this procedure retain scale-invariant geometry. More specifically, they possess geometrical properties of self-affine surfaces, with the scaling different in the (x, y) plane and the normal direction z.

To support the conjecture of fractal geometry of these samples, numerical modeling of the aggregation and deposition processes was made in [77]. In this model, an ensemble of three-dimensional lattice cluster-cluster aggregates is first generated according to the Meakin algorithm [7]; the details of this specific algorithm are described in Chap. 3. The fractal dimension of the generated clusters is close to value 1.8. Then the clusters are deposited onto a plain surface and allowed to undergo a vertical collapse so that the monomers that had an empty space directly beneath them are allowed to fall down until they hit the surface plane or another monomer. As a result, there are no empty spaces underneath monomers. Naturally, this projection from three to quasi-two dimensions results in a drastic restructuring of the clusters. An image of a single cluster-cluster aggregate with $N = 10^4$ monomers deposited on a surface is shown in Fig. 4.12a [77]. The calculated height-height correlation function $g(R)$ for the deposited samples has two characteristic correlation lengths: the first (short) length, l_1, is probably associated with random noise; the second (large) length, l_2, is of the order of the overall sample size. When R changes from 0 to l_1, the correlation function grows exponentially and saturates at some constant level, c. For $l_1 < R < l_2$, the function $[g(r) - c]$ grows according to the power law, as illustrated in Fig. 4.12b [77]. The scaling region extends from 30 to 400 lattice units and is well manifested. The corresponding fractal dimension of the samples, determined with the use of (4.1), is close to 2.6 in value.

4.4.2 Fractal-Surface-Enhanced Nonlinear Nano-Optics

Given all particle co-ordinates in a computer-simulated fractal aggregate deposited on a plane, we can solve the coupled-dipole equation CDE (3.2) for the general case, i.e. beyond the quasi-static approximation, by applying (for example) the conjugate gradient method. With the all local dipole moments found from the CDE, we can find the local field distribution and the spatial distributions of the local enhancement for various nonlinear optical processes and its averages.

As an example, in Figs. 4.13a and 4.13b the spatial distributions of local enhancements for second-harmonic generation at $\lambda = 1\,\mu\text{m}$ ($g_{\text{SHG}}(\mathbf{r}) = |E(\mathbf{r})/E_0|^4$; we set $E^{(0)} \equiv E_0$) and third-harmonic generation at $\lambda = 760\,\text{nm}$ ($g_{\text{THG}}(\mathbf{r}) = |E(\mathbf{r})/E_0|^6$) are shown. (In both cases interactions of the nonlinear dipoles at generated frequencies ω_g are neglected for the sake of simplicity.) We can see that most of the signal comes out of very small parts of the cluster (hot spots), where enhancement can be enormously large - much larger than the average enhancement for the sample. As mentioned, this opens a fascinating possibility of local nonlinear nano-spectroscopy of single molecules and nanoparticles.

We also note that the averaging of the distribution shown in Fig. 4.13 gives the mean enhancement for Raman scattering since $G_{RS} = \langle |E(\mathbf{r})/E_0|^4 \rangle$.

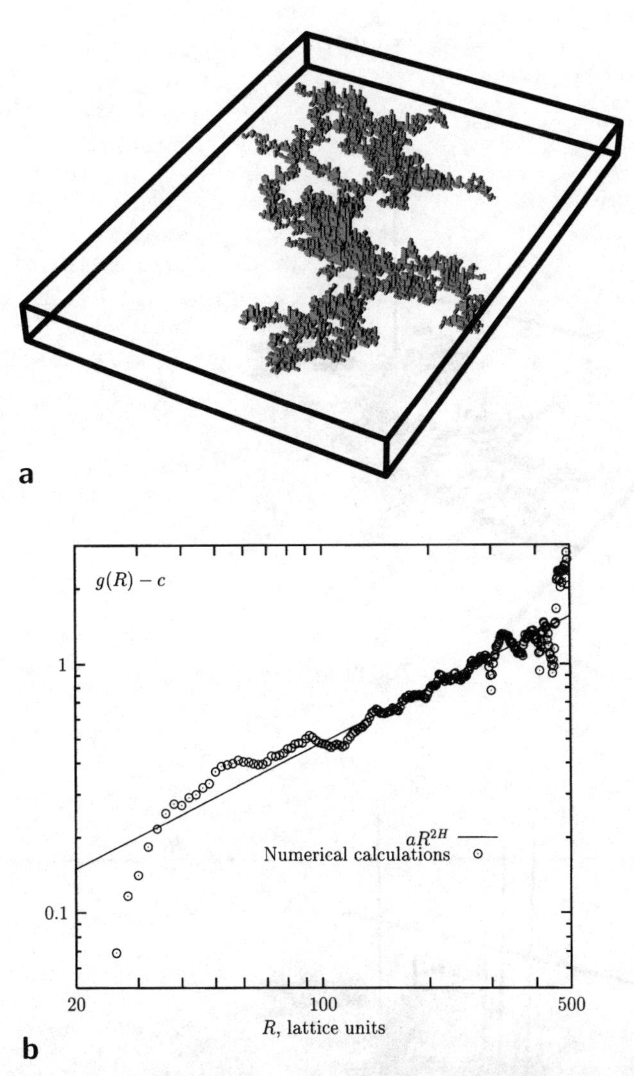

Fig. 4.12. (**a**) A computer-generated cluster-cluster aggregate deposited on a surface containing 10 000 elementary blocks. (**b**) The height-height correlation function $g(r)$ plotted *vs* the horizontal distance

In Fig. 4.14, the average enhancement G'_{K} for the optical Kerr effect (i.e. for the $\chi^{(3)}$ responsible for the nonlinear refraction) is shown. Note that G'_{K} is negative, so that the magnitudes $|G'_{\mathrm{K}}|$ are shown in Fig. 4.14. The enhancement strongly increases toward the infrared part of the spectrum reaching values up to 10^{10} at $\lambda \approx 10\,\mu\mathrm{m}$, whereas the average enhancement at $\lambda \approx 1\,\mu\mathrm{m}$ is $\sim 10^6$. It is important to stress again that the local enhancement in the hot spots can be much larger than the average. For instance, at

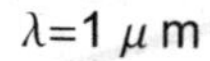

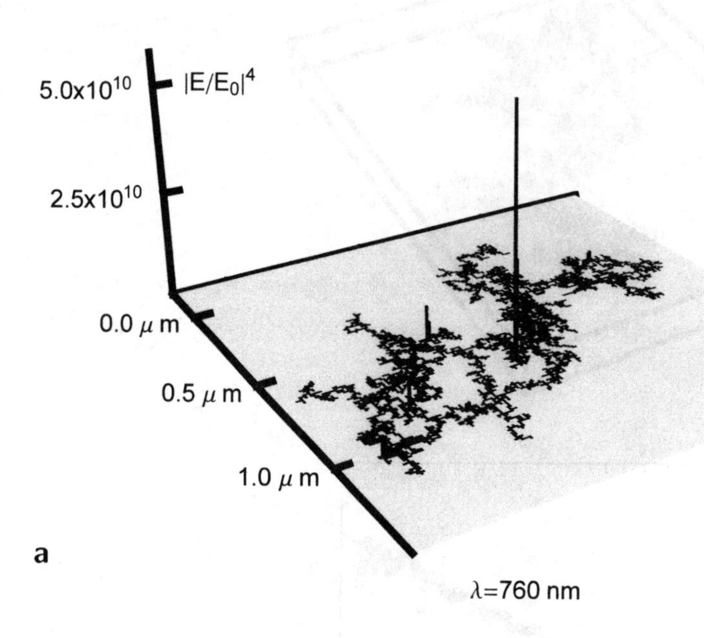

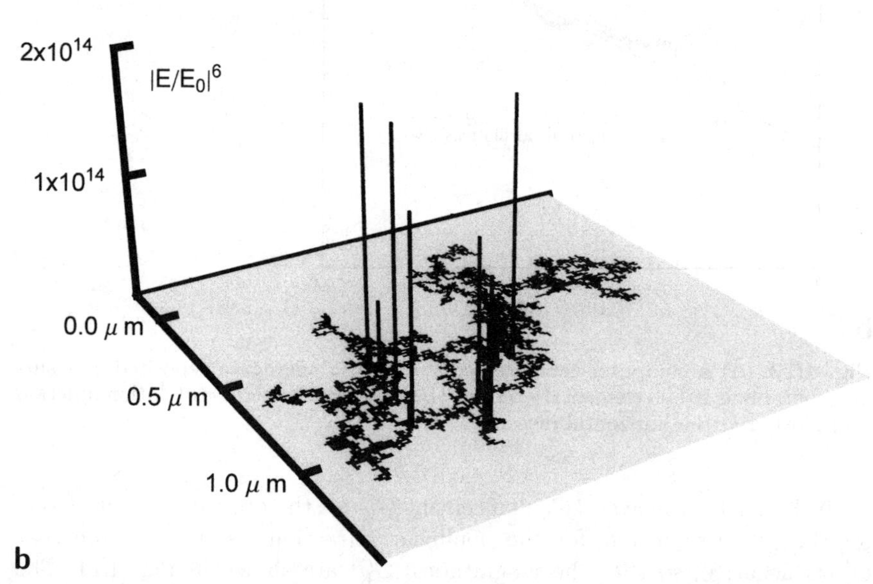

Fig. 4.13. Spatial distributions of local enhancements for (**a**) Second-harmonic generation at $\lambda = 1\,\mu\text{m}$ ($g_{\text{SHG}}(\mathbf{r}) = |E(\mathbf{r})/E_0|^4$) and (**b**) third-harmonic generation at $\lambda = 760\,\text{nm}$ ($g_{\text{THG}}(\mathbf{r}) = |E(\mathbf{r})/E_0|^6$). Here $E_0 \equiv E^{(0)}$ is the applied field

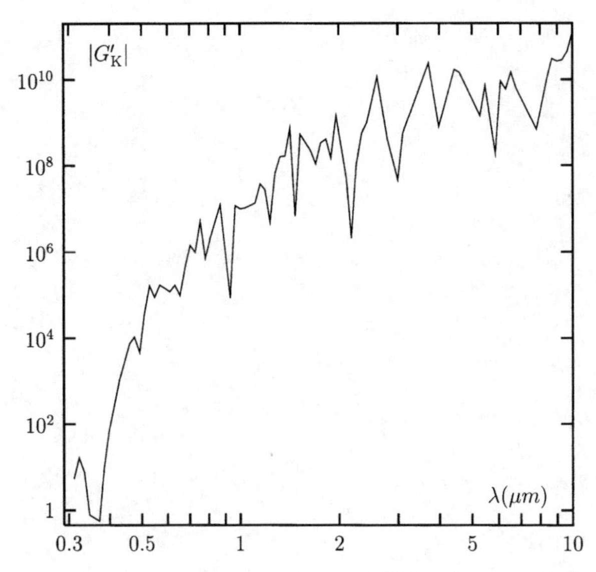

Fig. 4.14. Average enhancement G'_K for the optical Kerr effect in fractal aggregates deposited on a plane. The magnitudes of $G'_K < 0$ are shown

$\lambda \sim 1\,\mu m$, the local enhancement can be up to 10^{10}, which is four orders of magnitude larger than the average for the film.

Modern laser spectroscopy has already achieved extremely high resolution of 1 Hz and 5 fs in the spectral and time domains, respectively. By employing fractal thin films that increase drastically both the spatial resolution and the signal level, a new field, nonlinear optical nano-spectroscopy can be developed, which could become an unique tool for studying nanosized particles and single molecules.

5. Random Metal-Dielectric Films

Phenomena complex - Laws simple

Richard Feynman

5.1 Introduction

In this chapter we consider the optical properties of random metal-dielectric films, also referred to as semicontinuous metal films. These composite films form a large class of random media with high potential for various applications in optoelectronics.

Semicontinuous metal films can be produced by thermal evaporation or sputtering of metal onto an insulating substrate. In the growing process, first, small metallic grains are formed on the substrate. A typical size a of a metal grain is about 5 nm to 50 nm. As the film grows, the metal filling factor increases and coalescences occur, so that irregularly shaped clusters are formed on the substrate, eventually resulting in self-similar fractal structures. The concept of self-similarity plays an important role in the description of various properties of percolation systems [13, 26] (see also Sect. 1.4 in Chap. 1), and it will be used below in a scaling analysis of field fluctuations. The sizes of the fractal structures diverge in the vicinity of the percolation threshold, where an infinite cluster of metal is eventually formed, representing a continuous conducting path between the ends of a sample. At the percolation threshold the metal-insulator transition occurs in the system. At higher surface coverage, the film is mostly metallic, with voids of irregular shape. With further coverage increase, the film becomes uniform.

The optical properties of metal-dielectric films show anomalous phenomena that are absent for bulk metal and dielectric components. For example, the anomalous absorption in the near-infrared spectral range leads to unusual behavior of the transmittance and reflectance of the film. Typically, the transmittance is much higher than in continuous metal films, whereas the reflectance is much lower (see [13, 20, 26, 135–138]) and references therein). Near and well below the conductivity threshold, the anomalous absorptance can be as high as 50% [137, 139–141]. A number of effective-medium theories were proposed for calculation of the optical properties of random films, including the Maxwell-Garnett and Bruggeman approaches and their various modifications [20, 137, 138, 142]. The renormalization group method is also widely used to calculate effective dielectric response of two-dimensional percolating films near the percolation threshold (see [143, 144] and references therein). However, none of these theories allows one to calculate the field fluctuations and the effects resulting from these fluctuations.

A percolation system is very sensitive to the external electric field since its transport and optical properties are determined by a rather sparse network of conducting channels, and the field concentrates in the weak points of the channels. Therefore, percolation composite materials should have much larger nonlinear susceptibilities at zero and finite frequencies than its components.

Nonlinear electrical and optical properties of metal-dielectric percolation composites have attracted much attention in recent years. The distinguishing feature of random media (and, in particular, of percolation composites), namely the enhancement of nonlinearities in its components, was recognized very early on [45, 96, 97, 145, 146], and nonlinear conductivities and dielectric constants have been studied intensively in the last decade [96–101] (for a review of progress, see [11, 20]).

To avoid direct numerical calculations, the effective-medium theories, which have the virtue of relative mathematical and conceptual simplicity, were extended for the nonlinear response of percolation composites and fractal clusters. For linear problems, predictions of the effective-medium theory are usually sensible physically and offer quick insight into problems that are difficult to attack by other means [20]. The effective-medium theories, however, have disadvantages typical for all mean-field theories, namely, that they diminish fluctuations in a system. For example, these approaches assume that local electric fields are the same in the volume occupied by each component of a composite. The electric fields in different components are determined self-consistently. However, fluctuations in resonance random media are crucial in forming the nonlinear response, so that mean-field approaches are not justified in this case.

The local field fluctuations can be strongly enlarged, in the optical and infrared spectral ranges, for a composite material containing metal particles with negative real and small imaginary parts of the dielectric constant. In this case, the enhancement is due to the plasmon resonance in clusters of metallic granules [11, 20, 96]. The strong fluctuations of the local electric field lead to enhancement of various nonlinear effects. Nonlinear percolation composites are of practical importance as media with intensity-dependent dielectric functions and, in particular, as nonlinear optical switches, filters, and bistable elements.

In the remainder of this chapter, we consider in detail the field spatial distributions and nonlinear optical effects in random metal-dielectric films focusing on the visible and infrared spectral ranges where the plasmon resonances are effective in the films.

If the skin effect in metal grains is small, a semicontinuous film can be considered as a two-dimensional object. Then, in the optical spectral range where the frequency ω is much larger than the relaxation rate $\omega_\tau = \tau^{-1}$, a semicontinuous metal film can be thought of as a two-dimensional L–R–C lattice [20, 26]. The capacitance C stands for the gaps between metal grains that are filled by dielectric material (substrate) with the dielectric constant $\epsilon_{\rm d}$. The inductive elements L–R represent the metallic grains that for a Drude metal are characterized by the following dielectric function

$$\epsilon_{\rm m}(\omega) = \epsilon_0 - (\omega_{\rm p}/\omega)^2 \Big/ \left[1 + {\rm i}\omega_\tau/\omega\right], \tag{5.1}$$

where, as earlier, ϵ_0 is a contribution to $\epsilon_{\rm m}$ due to interband transitions, $\omega_{\rm p}$ is the plasma frequency, and $\omega_\tau = 1/\tau \ll \omega_{\rm p}$ is the relaxation rate (and it should be noted that in previous chapters we have also used the notation Γ for the relaxation rate). In the high-frequency range considered here, losses in metal grains are small, i.e. $\omega_\tau \ll \omega$. Therefore, the real part of the metal-dielectric function is much larger in magnitude than the imaginary part and it is negative for frequencies ω less than the "renormalized" plasma frequency defined as

$$\tilde{\omega}_{\rm p} = \omega_{\rm p}/\sqrt{\epsilon_0}. \tag{5.2}$$

Thus the metal conductivity is almost purely imaginary and metal grains can be, as mentioned, thought of as L–R elements, with the active component much smaller than the reactive one. Note that we use this L–R–C lattice representation only for illustrative purposes; the calculations below are general and do not actually rely on this representation.

It is instructive to consider first the film properties at the percolation threshold, $p = p_{\rm c}$, where the exact Dykhne's result for the effective dielectric constant $\epsilon_{\rm e} = \sqrt{\epsilon_{\rm d}\epsilon_{\rm m}}$ [147] holds in the quasistatic case. If we neglect metal losses and put $\omega_\tau = 0$, the metal-dielectric constant $\epsilon_{\rm m}$ is real and negative for frequencies smaller than the renormalized plasma frequency $\tilde{\omega}_{\rm p}$. We also neglect possible small losses in a dielectric substrate, assuming that $\epsilon_{\rm d}$ is real and positive. Then, $\epsilon_{\rm e}$ is purely imaginary for $\omega < \tilde{\omega}_{\rm p}$. Therefore, a film consisting of loss-free metal and dielectric grains is absorptive for $\omega < \tilde{\omega}_{\rm p}$. The effective absorption in a loss-free film means that the electromagnetic energy is stored in the system and thus the local fields are able to increase to an unlimited extent. In reality, the local fields in a metal film are, of course, finite because of the losses. If the losses are small, we can anticipate very strong field fluctuations. These large fluctuations result in giant enhancements of optical nonlinearities, as considered below.

This chapter is organized as follows. In Sect. 5.2 we briefly recapitulate the approach developed in [148]–[150] for calculating local fields in a semicontinuous film and define the enhancement factors of optical nonlinearities. We discuss the efficient numerical approach and results of calculations for local field distributions. We show that the local field distributions for both linear and nonlinear fields consist of very sharp field peaks, "hot" spots. The enhancement in these hot spots is enormous and exceeds a "background" nonlinear signal by many orders of magnitude. These novel effects can be obtained experimentally in the optical range by using, for example, near-field scanning optical microscopy allowing a sub-wavelength resolution. Section 5.3 describes a scaling theory for the field distributions and the high-order field moments that allow the finding of enhancements of different optical nonlinearities. In Sect. 5.4 we review some recent experimental results. Section 5.5 describes an interesting phenomenon, called percolation-enhanced nonlinear scattering (PENS), occurring in random metal-dielectric films.

In this chapter we use the "language" of currents and conductivities, instead of that of dipoles and polarizabilities used above. Although these two approaches are completely equivalent, historically the concept of currents and conductivities is typically used for the description of percolation films. This is because percolation films were originally studied in the low-frequency regime, where the Ohmic current prevails over the displacement field.

5.2 Giant Field Fluctuations

In metal-dielectric composites, the effective *dc conductivity* σ_{e} decreases with decreasing volume concentration p of the metal component and vanishes when the concentration p approaches the percolation threshold concentration p_{c}. In the vicinity of the percolation threshold p_{c}, the effective conductivity σ_{e} is determined by an infinite cluster of the percolating (conducting) channels. For concentrations p smaller than the percolation threshold p_{c}, the effective dc conductivity $\sigma_{\mathrm{e}} = 0$, that is the system is dielectric-like. Therefore metal–insulator transition takes place at $p = p_{\mathrm{c}}$. Since the metal-insulator transition associated with percolation represents a dynamic phase transition, the current and field fluctuations can be anticipated as scale-invariant and large.

In percolation composites, however, the fluctuation pattern appears to be quite different from that for a second-order transition, where fluctuations are characterized by a long-range correlation, with the relative magnitudes being of the order of unity at any point [155, 156]. In contrast, for *dc percolation*, the local electric fields are concentrated at the edges of large metal clusters, so that the field maxima (large fluctuations) are separated by distances of the order of the percolation correlation length ξ, which diverges when the metal volume concentration p approaches the percolation threshold p_{c}. We show below that the difference in fluctuations becomes even more striking in the optical spectral range, where the local field peaks have the resonance nature and, therefore, their relative magnitudes can be up to 10^5 for the linear response and up to 10^{20} for the nonlinear responses (e.g., for third-order optical nonlinearity), with distances between the peaks that can be much larger than ξ.

5.2.1 Linear Response

We consider here the optical properties of a semicontinuous film consisting of metal grains randomly distributed on a dielectric substrate. The film is placed in the (x, y) plane, whereas the incident wave propagates in the z direction. The local conductivity $\sigma(\mathbf{r})$ of the film takes either the "metallic" values $\sigma(\mathbf{r}) = \sigma_{\mathrm{m}}$ in metallic grains or the "dielectric" values $\sigma(\mathbf{r}) = -\mathrm{i}\omega\epsilon_{\mathrm{d}}/4\pi$ outside the metallic grains. The vector $\mathbf{r} = (x, y)$ has two components in the plane of the film, and ω is the frequency of the incident wave. The gaps between metallic grains are assumed to be filled by the material of the substrate, so that the ϵ_d introduced above is assumed equal to the dielectric constant of the substrate. For a two-dimensional problem, the electric field in the film is assumed to be homogeneous in the direction z perpendicular to the film plane; this means that the skin depth for the metal, $\delta \cong c/(\omega\sqrt{|\epsilon_{\mathrm{m}}|})$, is much larger than the metal grain size a and the quasistatic approximation can be applied for calculating the field distributions. We also take into account that the wavelength of the incident wave is much larger than any characteristic size of the film, including the grain size and the gaps between the grains. In this case, the local field $\mathbf{E}(\mathbf{r})$ can be represented as

$$\mathbf{E}(\mathbf{r}) = -\nabla\phi(\mathbf{r}) + \mathbf{E}_{\mathrm{e}}(\mathbf{r}), \tag{5.3}$$

where $\mathbf{E}_{\mathrm{e}}(\mathbf{r})$ is the applied (macroscopic) field and $\phi(\mathbf{r})$ is the potential of the fluctuating field inside the film. The current density $\mathbf{j}(\mathbf{r})$ at point $\mathbf{r}$ is given by Ohm's law,

$$\mathbf{j}(\mathbf{r}) = \sigma(\mathbf{r})\left[-\nabla\phi(\mathbf{r}) + \mathbf{E}_{\mathrm{e}}(\mathbf{r})\right]. \tag{5.4}$$

The current conservation law, $\nabla \cdot \mathbf{j}(\mathbf{r}) = 0$, has the following form

$$\nabla \cdot \left(\sigma(\mathbf{r})\left[-\nabla\phi(\mathbf{r}) + \mathbf{E}_{\mathrm{e}}(\mathbf{r})\right]\right) = 0. \tag{5.5}$$

We solve (5.5) to find the fluctuating potential $\phi(\mathbf{r})$ and the local field $\mathbf{E}(\mathbf{r})$ induced in the film by the applied field $\mathbf{E}_{\mathrm{e}}(\mathbf{r})$. When the wavelength of the incident em-wave is much larger than all spatial scales of a semicontinuous metal film, the applied field $\mathbf{E}_{\mathrm{e}}$, i.e. the field of the incident wave, is constant in the film plane: $\mathbf{E}_{\mathrm{e}}(\mathbf{r}) = \mathbf{E}^{(0)}$. Provided that the local field $\mathbf{E}(\mathbf{r})$ is known, the effective conductivity σ_{e} can be obtained from the definition

$$\langle\mathbf{j}(\mathbf{r})\rangle = \sigma_{\mathrm{e}}\mathbf{E}^{(0)}, \tag{5.6}$$

where the symbol $\langle\cdots\rangle$ denotes the average over the entire film.

The local field $\mathbf{E}(\mathbf{r})$, induced by the applied field $\mathbf{E}_{\mathrm{e}}(\mathbf{r})$ can be obtained by using the nonlocal conductivity $\hat{\Sigma}$ introduced in [149]:

$$\mathbf{E}(\mathbf{r}) = \frac{\mathbf{j}(\mathbf{r})}{\sigma(\mathbf{r})} = \frac{1}{\sigma(\mathbf{r})}\int \hat{\Sigma}(\mathbf{r}, \mathbf{r}')\mathbf{E}_{\mathrm{e}}(\mathbf{r}')\, d\mathbf{r}'. \tag{5.7}$$

According to (5.7), the nonlocal conductivity $\hat{\Sigma}(\mathbf{r}, \mathbf{r}')$ relates the applied field at point $\mathbf{r}'$ to the current and the local field at point $\mathbf{r}$. The nonlocal conductivity in (5.7) can be expressed in terms of the Green function of the equation (5.5):

$$\Sigma_{\alpha\beta}(\mathbf{r}_2, \mathbf{r}_1) = \sigma(\mathbf{r}_2)\sigma(\mathbf{r}_1)\frac{\partial^2 G(\mathbf{r}_2, \mathbf{r}_1)}{\partial r_{2\,\alpha}\, \partial r_{1\,\beta}}, \tag{5.8}$$

where the Greek indices represent the x and y components. The Green function is symmetrical with respect to the interchange of its arguments: $G(\mathbf{r}_1, \mathbf{r}_2) = G(\mathbf{r}_2, \mathbf{r}_1)$; thus, (5.8) implies that the nonlocal conductivity is also symmetrical:

$$\Sigma_{\alpha\beta}(\mathbf{r}_1, \mathbf{r}_2) = \Sigma_{\beta\alpha}(\mathbf{r}_2, \mathbf{r}_1). \tag{5.9}$$

The nonlocal conductivity $\hat{\Sigma}$, introduced above, is useful for analysis of different optical processes in the system.

For further consideration of nonlinear effects, we suppose now that the applied field has the following form:

$$\mathbf{E}_{\mathrm{e}}(\mathbf{r}) = \mathbf{E}^{(0)} + \mathbf{E}_{\mathrm{f}}(\mathbf{r}), \tag{5.10}$$

where $\mathbf{E}^{(0)}$ is the constant linearly polarized field, and fluctuating field $\mathbf{E}_{\mathrm{f}}(\mathbf{r})$ may arbitrarily change over the film surface but its averaged value $\langle \mathbf{E}_{\mathrm{f}}(\mathbf{r}) \rangle$ is collinear to $\mathbf{E}^{(0)}$. Then the average current density $\langle \mathbf{j} \rangle$ is also collinear to $\mathbf{E}^{(0)}$ in the macroscopically isotropic films considered here. Therefore, the average current can be written as

$$\langle \mathbf{j}(\mathbf{r}) \rangle = \frac{\mathbf{E}^{(0)}}{E^{(0)2}}\left(\mathbf{E}^{(0)} \cdot \langle \mathbf{j}(\mathbf{r}) \rangle\right) = \frac{\mathbf{E}^{(0)}}{E^{(0)2}}\frac{1}{A}\int E_{\alpha}^{(0)} j_{\alpha}(\mathbf{r})\, \mathrm{d}\mathbf{r}, \tag{5.11}$$

where A is the total area of the film, the integration is over the film area and $E^{(0)2} \equiv \left(\mathbf{E}^{(0)} \cdot \mathbf{E}^{(0)}\right)$. By expressing the current $j_{\alpha}(\mathbf{r})$ in (5.11) in terms of the nonlocal conductivity matrix $j_{\alpha}(\mathbf{r}) = \int \Sigma_{\alpha\beta}(\mathbf{r}, \mathbf{r}_1) E_{\mathrm{e}\,\beta}(\mathbf{r}_1)\, \mathrm{d}\mathbf{r}_1$, we obtain

$$\langle \mathbf{j}(\mathbf{r}) \rangle = \frac{\mathbf{E}^{(0)}}{E^{(0)2}}\frac{1}{A}\int E_{\alpha}^{(0)} \Sigma_{\alpha\beta}(\mathbf{r}, \mathbf{r}_1) E_{\mathrm{e}\,\beta}(\mathbf{r}_1)\, \mathrm{d}\mathbf{r}\, \mathrm{d}\mathbf{r}_1, \tag{5.12}$$

where the integrations are over the entire film. We can integrate this equation over the co-ordinates $\mathbf{r}$ and use the symmetry of the matrix of nonlocal conductivity given by (5.9), which results in the formula

$$\langle \mathbf{j}(\mathbf{r}) \rangle = \frac{\mathbf{E}^{(0)}}{E^{(0)2}}\frac{1}{A}\int j_{0\,\beta}(\mathbf{r}_1) E_{\mathrm{e}\,\beta}(\mathbf{r}_1)\, \mathrm{d}\mathbf{r}_1, \tag{5.13}$$

where $\mathbf{j}_0(\mathbf{r})$ is the current induced at the co-ordinate $\mathbf{r}$ by the constant external field $\mathbf{E}^{(0)}$. Now we can substitute in (5.13) the external field $\mathbf{E}_{\mathrm{e}}(\mathbf{r})$ from (5.10) and integrate over the co-ordinate $\mathbf{r}_1$, which gives us

$$\langle \mathbf{j}(\mathbf{r}) \rangle = \frac{\mathbf{E}^{(0)}}{E^{(0)2}}\left(\mathbf{E}^{(0)} \cdot \langle \mathbf{j}_0(\mathbf{r}) \rangle\right) + \frac{\mathbf{E}^{(0)}}{E^{(0)2}}\left\langle (\sigma(\mathbf{r})\mathbf{E}(\mathbf{r}) \cdot \mathbf{E}_{\mathrm{f}}(\mathbf{r})) \right\rangle, \tag{5.14}$$

where the field $\mathbf{E}(\mathbf{r}) = \mathbf{j}_0(\mathbf{r})/\sigma(\mathbf{r})$ is the local field induced in the film by the constant external field $\mathbf{E}^{(0)}$. Substituting in (5.14) the expression for the effective conductivity given by (5.6), we obtain the following equation:

$$\langle \mathbf{j}(\mathbf{r}) \rangle = \mathbf{E}^{(0)} \left[\sigma_{\mathrm{e}} + \frac{\langle \sigma(\mathbf{r}) \left(\mathbf{E}(\mathbf{r}) \cdot \mathbf{E}_{\mathrm{f}}(\mathbf{r}) \right) \rangle}{E^{(0)2}} \right]. \tag{5.15}$$

Thus, the average current induced in a macroscopically isotropic film by a nonuniform external field $\mathbf{E}_{\mathrm{e}}(\mathbf{r})$ can be expressed in terms of the fluctuating part $\mathbf{E}_{\mathrm{f}}(\mathbf{r})$ of the external field and the local field induced in the film by the constant part $\mathbf{E}^{(0)}$ of the external field.

Below we shall use (5.15) in an analysis of the nonlinear response of semicontinuous metal films, since it allows us to express various nonlinear currents in terms of the local fields.

5.2.2 Nonlinear Response

Consider now a composite with local conductivity $\sigma(\mathbf{r})$ including the cubic nonlinearity, i.e., $\sigma(\mathbf{r}) = \sigma^{(1)}(\mathbf{r}) + \sigma^{(3)}(\mathbf{r})|\mathbf{E}(\mathbf{r})|^2$. To find the effective conductivity σ_{e} (which, of course, is also nonlinear), we write (5.5) in the following form

$$\nabla \cdot \left(\sigma^{(1)}(\mathbf{r}) \left[-\nabla\phi(\mathbf{r}) + \mathbf{E}^{(0)} + \frac{\sigma^{(3)}(\mathbf{r})}{\sigma^{(1)}(\mathbf{r})} \mathbf{E}'(\mathbf{r}) |\mathbf{E}'(\mathbf{r})|^2 \right] \right) = 0, \tag{5.16}$$

where $\mathbf{E}'(\mathbf{r}) = -\nabla\phi(\mathbf{r}) + \mathbf{E}^{(0)}$ is the local electric field at the co-ordinate $\mathbf{r}$ in the nonlinear film, with the local conductivity containing the cubic term. We treat the last term in the square brackets as a fluctuating "external" inhomogeneous field ($\mathbf{E}_f(\mathbf{r})$) and use (5.15) to obtain the average current

$$\langle \mathbf{j}(\mathbf{r}) \rangle = \sigma_{\mathrm{e}}^{(1)} \mathbf{E}^{(0)} + \mathbf{E}^{(0)} \frac{\left\langle \sigma^{(3)}(\mathbf{r}) \left(\mathbf{E}(\mathbf{r}) \cdot \mathbf{E}'(\mathbf{r}) \right) |\mathbf{E}'(\mathbf{r})|^2 \right\rangle}{E^{(0)2}}, \tag{5.17}$$

where $\sigma_{\mathrm{e}}^{(1)}$ is the linear effective conductivity and $\mathbf{E}(\mathbf{r})$ is the local field found in the linear approximation (i.e., for the local conductivity, $\sigma(\mathbf{r}) \equiv \sigma^{(1)}(\mathbf{r})$). Equation (5.17) expresses the average current and thus the effective nonlinear conductivity, in terms of the local fields $\mathbf{E}'(\mathbf{r})$ and $\mathbf{E}(\mathbf{r})$.

For a relatively weak nonlinearity, when $\sigma^{(3)}(\mathbf{r})|\mathbf{E}(\mathbf{r})|^2 \ll \sigma^{(1)}(\mathbf{r})$, we can replace the local field $\mathbf{E}'(\mathbf{r})$ in (5.17) by the field $\mathbf{E}(\mathbf{r})$ calculated in the linear approximation. This gives

$$\langle \mathbf{j}(\mathbf{r}) \rangle = \left(\sigma_{\mathrm{e}}^{(1)} + \frac{\langle \sigma^{(3)}(\mathbf{r}) E^2(\mathbf{r}) |\mathbf{E}(\mathbf{r})|^2 \rangle}{E^{(0)2}} \right) \mathbf{E}^{(0)}, \tag{5.18}$$

where $E^2(\mathbf{r}) \equiv (\mathbf{E}(\mathbf{r}) \cdot \mathbf{E}(\mathbf{r}))$. From this equation, it follows that the effective nonlinear conductivity σ_{e} has the form

$$\sigma_{\mathrm{e}} = \sigma_{\mathrm{e}}^{(1)} + \sigma_{\mathrm{e}}^{(3)} |\mathbf{E}^{(0)}|^2, \tag{5.19}$$

where the effective nonlinear conductivity $\sigma_{\mathrm{e}}^{(3)}$ is given by

$$\sigma_{\mathrm{e}}^{(3)} = \frac{\left\langle \sigma^{(3)}(\mathbf{r}) E^2(\mathbf{r}) |\mathbf{E}(\mathbf{r})|^2 \right\rangle}{E^{(0)2} \left|\mathbf{E}^{(0)}\right|^2}. \tag{5.20}$$

Equation (5.20) thus expresses $\sigma_{\mathrm{e}}^{(3)}$ in terms of the local fields $\mathbf{E}(\mathbf{r})$ obtained in the linear approximation.

When the local fields in a system experience large fluctuations, the effective nonlinearity $\sigma_{\mathrm{e}}^{(3)}$ is strongly enhanced in comparison with the average $\langle \sigma^{(3)}(\mathbf{r}) \rangle$.

Now let us suppose for simplicity that the conductivities of the adsorbed molecules $\sigma^{(3)}$ are uniformly distributed over the film surface. Then (5.20) simplifies to

$$\sigma_{\mathrm{e}}^{(3)} = \sigma^{(3)} \frac{\left\langle E^2(\mathbf{r}) |\mathbf{E}(\mathbf{r})|^2 \right\rangle}{E^{(0)2} \left|\mathbf{E}^{(0)}\right|^2}. \tag{5.21}$$

In the absence of metal grains, the effective nonlinear conductivity $\sigma_{\mathrm{e}}^{(3)}$ coincides with the conductivity $\sigma^{(3)}$ of the layer of the absorbed nonlinear molecules.

By introducing

$$G_{\mathrm{K}} = \frac{\left\langle E^2(\mathbf{r}) |\mathbf{E}(\mathbf{r})|^2 \right\rangle}{E^{(0)2} \left|\mathbf{E}^{(0)}\right|^2}, \tag{5.22}$$

we can write the current in the form $\mathbf{j}(\mathbf{r}) = \mathbf{j}^{(1)}(\mathbf{r}) + \mathbf{j}^{(3)}(\mathbf{r})$, where $\mathbf{j}^{(1)}(\mathbf{r})$ is the linear current and $\mathbf{j}^{(3)}(\mathbf{r})$ is the nonlinear one that can be represented as

$$\left\langle \mathbf{j}^{(3)}(\mathbf{r}) \right\rangle = \sigma_{\mathrm{e}}^{(3)} \left|\mathbf{E}^{(0)}\right|^2 \mathbf{E}^{(0)}, \tag{5.23}$$

with

$$\sigma_{\mathrm{e}}^{(3)} = G_{\mathrm{K}} \sigma^{(3)}. \tag{5.24}$$

As discussed in the previous chapters, the Kerr-type optical nonlinearity is a third-order optical nonlinearity that results in the nonlinear polarization $\mathbf{P}^{(3)}$ which is given by

$$P_{\alpha}^{(3)}(\omega) = \chi_{\alpha\beta\gamma\delta}^{(3)}(-\omega; \omega, \omega, -\omega) E_{\beta} E_{\gamma} E_{\delta}^{*}, \tag{5.25}$$

where $\chi_{\alpha\beta\gamma\delta}^{(3)}(-\omega; \omega, \omega, -\omega)$. is the nonlinear susceptibility [55], and $\mathbf{E}$ is an electric field at frequency ω. Summation over repeated Greek indices is implied. The Kerr optical nonlinearity results in nonlinear corrections (proportional to the light intensity) in the refractive index and absorption coefficient.

In an isotropic medium with linear light polarization, the Kerr-type polarization has the simple form

$$\left\langle \mathbf{P}^{(3)}(\mathbf{r}) \right\rangle = \chi_{\mathrm{e}}^{(3)} \left| \mathbf{E}^{(0)} \right|^2 \mathbf{E}^{(0)}, \tag{5.26}$$

where the nonlinear susceptibility $\chi_{\mathrm{e}}^{(3)}$ is a scalar.

By comparing (5.23) and (5.26), we can conclude that $\sigma_{\mathrm{e}}^{(3)}$ is the Kerr-type nonlinear conductivity which is a complete analog of $\chi^{(3)}$ in the "language" of currents and conductivities. The effective Kerr conductivity is $\sigma_{\mathrm{e}}^{(3)} = -\mathrm{i}\omega\epsilon_{\mathrm{e}}^{(3)}/4\pi$, where $\epsilon_{\mathrm{e}}^{(3)} = 4\pi\chi_{\mathrm{e}}^{(3)}$. Accordingly, G_{K} in (5.22) defines the enhancement-factor of the Kerr nonlinearity resulting from placing Kerr-active molecules on the surface of a metal-dielectric film. We also note that the formula (5.22) coincides with formula (2.1) given in Chap. 2.

In the discussion above, we assumed that the nonlinear Kerr conductivity $\sigma^{(3)}$ is due to molecules covering a film. In some cases, the nonlinear response can be also due to the metal and/or dielectric grains forming the film, with no adsorbed molecules on it. If this is the case, then $\sigma^{(3)}(\mathbf{r})$ refers to the nonlinear conductivity of a grain. Repeating the above derivations, we arrive at the following result for the effective Kerr conductivity

$$\sigma_{\mathrm{e}}^{(3)} = p\sigma_{\mathrm{m}}^{(3)} \frac{\left\langle E^2(\mathbf{r})|\mathbf{E}(\mathbf{r})|^2 \right\rangle_m}{E^{(0)2}\left|\mathbf{E}^{(0)}\right|^2} + (1-p)\sigma_{\mathrm{d}}^{(3)} \frac{\left\langle E^2(\mathbf{r})|\mathbf{E}(\mathbf{r})|^2 \right\rangle_{\mathrm{d}}}{E^{(0)2}\left|\mathbf{E}^{(0)}\right|^2}, \tag{5.27}$$

where the angular brackets $\langle ... \rangle_{\mathrm{m}}$ and $\langle ... \rangle_{\mathrm{d}}$ stand for the averaging over the metal and dielectric grains respectively, and $\sigma_{\mathrm{m}}^{(3)}$ and $\sigma_{\mathrm{d}}^{(3)}$ are the corresponding nonlinear conductivities.

Note that $\mathbf{E}^{(0)}$ is actually the average macroscopic field, which can in general be different from the incident field $\mathbf{E}_{\mathrm{inc}}$. For thin two-dimensional films at the quasi-static limit, the macroscopic field is constant and related to the incident field through the transmittance T as $\mathbf{E}^{(0)} = T\mathbf{E}_{\mathrm{inc}}$. Above, we defined the enhancement factor as the ratio of the nonlinear signals from a film with and without metal grains on it. This means that in the denominator of the expression (5.22) we should in this case replace $[\mathbf{E}^{(0)}]^4$ by $[\mathbf{E}_{\mathrm{inc}}]^4 = [\mathbf{E}^{(0)}]^4/T^4$, which gives an additional pre-factor T^4 in (5.22) if by $\mathbf{E}^{(0)}$ we mean the macroscopic field. (For a purely dielectric film without metal grains, we can set $T_{\mathrm{d}} = 1$ and $\mathbf{E}^{(0)} = \mathbf{E}_{\mathrm{inc}}$.) For the sake of simplicity, hereafter we omit this pre-factor associated with the transmittance T. To take it into account, we should make the above replacement $\mathbf{E}^{(0)} \to \mathbf{E}^{(0)}/T$ in the denominators of formulas for enhancements of nonlinear optical processes.

As shown in Chap. 2, enhancement for nearly-degenerate four-wave mixing [compare (2.2)] is given by

$$G_{\mathrm{FWM}} = |G_{\mathrm{K}}|^2. \tag{5.28}$$

Similarly, we can find enhancement for Raman scattering from molecules on the surface of a metal-dielectric film [149], where

$$G_{\mathrm{RS}} = \frac{\left\langle |\epsilon(\mathbf{r})|^2 |\mathbf{E}(\mathbf{r})|^4 \right\rangle}{|\epsilon_{\mathrm{d}}|^2 \left|\mathbf{E}^{(0)}\right|^4}. \tag{5.29}$$

Note that (5.29) differs from (2.3) by a factor of $|\sigma(\mathbf{r})/\sigma_{\mathrm{d}}||^2 = |\epsilon(\mathbf{r})/\epsilon_{\mathrm{d}}|^2$. The difference results from the fact that in (2.3) we assumed that both the Raman and linear polarizabilities are associated with the same site on a fractal surface, whereas to obtain (5.29) it was assumed that the linear polarizability is due to a metal grain on the film but the Raman polarizability is due to a molecule adsorbed on the grain. However, as shown below, the largest fields are concentrated in dielectric gaps where $\epsilon(\mathbf{r}) = \epsilon_{\mathrm{d}}$, so that we can use the formula

$$G_{\mathrm{RS}} = \frac{\left\langle |\mathbf{E}(\mathbf{r})|^4 \right\rangle}{\left|\mathbf{E}^{(0)}\right|^4}. \tag{5.30}$$

Finally, for third-harmonic generation we can obtain the following formula (cf. (2.5); for details see [150]):

$$G_{\mathrm{THG}} = \left| \frac{\left\langle \mathbf{j}^{(3)}_{3\omega}(\mathbf{r}) \right\rangle}{\left\langle \mathbf{j}^{(3)}_{3\omega\,0}(\mathbf{r}) \right\rangle} \right|^2 = \left| \frac{\left\langle \epsilon_{3\omega}(\mathbf{r}) \left(\mathbf{E}_{3\omega}(\mathbf{r}) \cdot \mathbf{E}_{\omega}(\mathbf{r})\right) E^2_{\omega}(\mathbf{r}) \right\rangle}{\epsilon_{\mathrm{d}} \left(\mathbf{E}^{(0)}_{3\omega} \cdot \mathbf{E}^{(0)}_{\omega}\right) {E^{(0)}_{\omega}}^2} \right|^2. \tag{5.31}$$

By omitting again for simplicity the factor $\epsilon(\mathbf{r})/\epsilon_{\mathrm{d}}$, we obtain

$$G_{\mathrm{THG}} = \left| \frac{\left\langle \left(\mathbf{E}_{3\omega}(\mathbf{r}) \cdot \mathbf{E}_{\omega}(\mathbf{r})\right) E^2_{\omega}(\mathbf{r}) \right\rangle}{\left(\mathbf{E}^{(0)}_{3\omega} \cdot \mathbf{E}^{(0)}_{\omega}\right) {E^{(0)}_{\omega}}^2} \right|^2. \tag{5.32}$$

In the case when the generated field $\mathbf{E}_{\omega}(\mathbf{r})$ does not excite the plasmon resonances in the film and the 3ω field is uniform, so that $\mathbf{E}_{3\omega}(\mathbf{r}) = \mathbf{E}^{(0)}_{3\omega}$, the local currents $\mathbf{j}^{(3)}_{3\omega}(\mathbf{r})$ and $\mathbf{j}^{(3)}_{3\omega\,0}(\mathbf{r})$ depend only on the local conductivities and fields at point $\mathbf{r}$. As shown below, the distribution of $\left|\mathbf{j}^{(3)}_{3\omega}(\mathbf{r})\right|^2 \propto \left|\mathbf{E}_{\omega}(\mathbf{r})\, {E_{\omega}}^2(\mathbf{r})\right|^2$ consists of spatially separated large peaks that can be probed independently by means of near-field scanning optical microscopy (NSOM). This means that in this case we can define the spatial distribution of the local third-harmonic signals $I_{3\omega}(\mathbf{r}) \propto \left|\mathbf{j}^{(3)}_{3\omega}(\mathbf{r})\right|^2$, and the local enhancements for THG are defined as follows:

$$g_{\mathrm{THG}}(\mathbf{r}) = \left| \frac{\mathbf{j}^{(3)}_{3\omega}(\mathbf{r})}{\mathbf{j}^{(3)}_{3\omega\,0}(\mathbf{r})} \right|^2 = \left| \frac{\mathbf{E}_{\omega}(\mathbf{r})\, {E_{\omega}}^2(\mathbf{r})}{\mathbf{E}^{(0)}_{\omega}\, {E^{(0)}_{\omega}}^2} \right|^2. \tag{5.33}$$

Similar formulas can be obtained for the surface-enhanced second harmonic generation (SHG) [150].

Thus, enhancement of various nonlinear optical processes within a certain approximation can be expressed in terms of the local fields, as shown above. To calculate the local electric fields in two-dimensional systems, we discretize (5.5) on a square lattice. The potentials in the sites of the lattice reproduce the local field potentials in a semicontinuous film. The conductivities of the lattice bonds stand for the local film conductivity and take either σ_{m} or σ_{d} values. In such a way, the partial differential equation (5.5) is reduced to a set of Kirchhoff's equations that can be solved by the method discussed in the next section. Provided that the field distribution is known, we can use formulas such as (5.22) and others to calculate enhancements of the optical nonlinearities. Below, in the remainder of this section, we first consider the numerical method and then the field distributions.

5.2.3 Numerical Model

There exist now very efficient numerical methods for calculating the effective conductivity of composite materials [20, 26, 27, 28], but they typically do not allow calculations of the field distributions. Here, we consider the real space renormalization group (RSRG) method that was suggested by Reynolds et al.[151] and Sarychev [152] and then extended to study the conductivity [153] and permeability of oil reservoirs [154]. The approach used by Aharony [154] can be adopted for finding the field distributions in the following way [149, 150].

First, we generate a square lattice of L–R (metal) and C (dielectric) bonds using a random-number generator. As seen in Fig. 5.1, such a lattice can be considered as a set of "corner" elements. One such element is labeled as (ABCDEFGH) in Fig. 5.1. In the first stage of the RSRG procedure, each of these elements is replaced by two Wheatstone bridges, as shown in Fig. 5.1. After this transformation, the initial square lattice is converted to another square lattice, with the distance between the sites twice larger and with

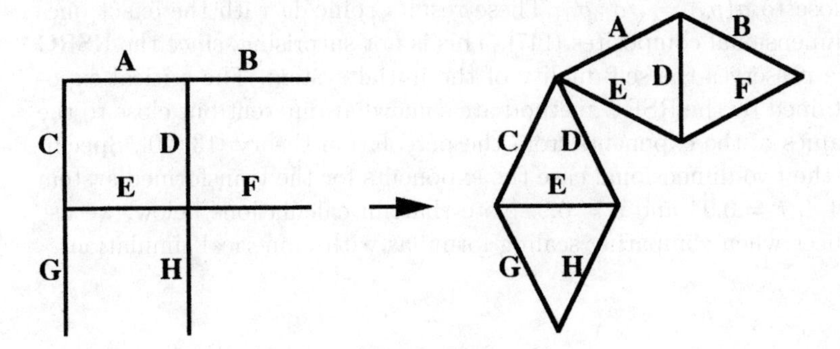

Fig. 5.1. Real space renormalization scheme

each bond between the two nearest neighboring sites being the Wheatstone bridge. Note that there is a one-to-one correspondence between the "x" bonds in the initial lattice and the "x" bonds in the "x"-directed bridges of the transformed lattice, as seen in Fig. 5.1. The same one-to-one correspondence exists also between the "y" bonds. The transformed lattice is also a square lattice, and we can again apply to it the RSRG transformation. We continue this procedure until the size l of the system is reached. As a result, instead of the initial lattice we have two large Wheatstone bridges in the "x" and "y" directions. Each of them has a hierarchical structure consisting of bridges with sizes from 2 to l. Because the one-to-one correspondence is preserved at each step of the transformation, the correspondence also exists between the elementary bonds of the transformed lattice and the bonds of the initial lattice.

After using the RSRG transformation, we apply an external field to the system and solve Kirchhoff's equations to determine the fields and the currents in all the bonds of the transformed lattice. Due to the hierarchical structure of the transformed lattice, these equations can be solved exactly. Then, we use the one-to-one correspondence between the elementary bonds of the transformed lattice and the bonds of the initial square lattice to find the field distributions in the initial lattice as well as its effective conductivity. The number of operations to get the full distributions of the local fields is proportional to l^2, compared with l^7 operations needed in the transform-matrix method [20] and l^3 operations needed in the well known Frank-Lobb algorithm [28], which does not provide the field distributions but the effective conductivity only.

However, the RSRG procedure is not exact, since the effective connectivity of the transformed system does not repeat exactly the connectivity of the initial square lattice. To check the accuracy of the RSRG result, the two-dimensional percolation problem has been solved using this method [148] to find the effective parameters of a two-component composite with the real metallic conductivity σ_{m} much larger than the real conductivity σ_{d} of the dielectric component, i.e. $\sigma_{\mathrm{m}} \gg \sigma_{\mathrm{d}}$. It was obtained that a percolation threshold occurs at $p_{\mathrm{c}} = 0.5$ and the effective conductivity at the percolation threshold is very close to $\sigma(p_{\mathrm{c}}) = \sqrt{\sigma_{\mathrm{m}}\sigma_{\mathrm{d}}}$. These results coincide with the exact ones for two-dimensional composites [147]. This is not surprising, since the RSRG procedure preserves the self-duality of the initial system. The critical exponents obtained by the RSRG method are somewhat different but close to the known values of the exponents from the percolation theory [13, 20]. Specifically, for the two-dimensional case the exponents for the transformed system are $\nu = 1.4$, $t = 0.94$ and $s = 0.9$. Note that, in calculations below, we use these indices when comparing scaling formulas with numerical simulations.

5.2.4 Field Distributions

In this section we discuss the results of numerical simulations of local fields using the method described above. The applied field $E^{(0)}$ is set to unity, whereas the local fields inside the system are complex quantities. The dielectric constant of silver grains, for example, has the form of (5.1) with an interband-transitions contribution $\epsilon_0 = 5.0$, plasma frequency $\omega_{\mathrm{p}} = 9.1$ eV, and relaxation frequency $\omega_\tau = 0.021$ eV. Below we set $\epsilon_{\mathrm{d}} = 2.2$, which is typical for glass.

Figure 5.2 shows the field distributions $G(\mathbf{r}) = \left|E(\mathbf{r})/E^{(0)}\right|^2$ for the plasmon resonance frequency $\omega = \omega_{\mathrm{r}}$ that in two-dimensional systems corresponds to the condition $\mathrm{Re}(\epsilon_{\mathrm{m}}(\omega_{\mathrm{r}})) = -\epsilon_{\mathrm{d}}$. The value of the frequency ω_{r} is only slightly below the renormalized plasma frequency $\tilde{\omega}_{\mathrm{p}}$ defined above in (5.2). For silver particles, the resonance condition is fulfilled at wavelength $\lambda \approx 365$ nm. The frequency ω_{r} gives the resonance of an isolated metal particle. (For the z-independent two-dimensional problem, particles can be thought of as cylinders infinite in the z-direction that in the quasistatic approximation resonate at the frequency $\omega = \omega_{\mathrm{r}}$, for the field polarized in the x, y plane). The results are presented for various metal concentrations p.

For $p = 0.001$ (Fig. 5.2a), metal grains practically do not interact, so that all the peaks are almost of the same height and indicate the locations of metal particles. Note that a similar distribution is obtained for $p = 0.999$ (Fig. 5.2g) when the role of metal particles is played by dielectric voids. For $p = 0.1$ (Fig. 5.2c) and especially for $p = 0.5$ (Fig. 5.2d), metal grains form clusters of strongly interacting particles. These clusters resonate at frequencies different from an isolated particle; therefore, for the chosen frequency $\omega = \omega_{\mathrm{r}}$, the field peaks are smaller on average than those for the individual particles, and the height distribution is very inhomogeneous. Note that the spatial scale for the local field distribution is much larger than the metal grain size a (chosen to be unity in all figures).

As follows from Fig. 5.2 (see also Fig. 5.3), the main assumption of the effective-medium theory, that the local fields are the same for all metal grains, fails for the plasmon resonance frequencies and nonvanishing concentrations p. We also emphasize a strong resemblance in the field distributions for p and $1 - p$ (compare Fig. 5.2a and g, b and f, c and e).

For larger wavelengths, a single metal grain is off the plasmon resonance. Nevertheless, as one can see from Figs. 5.3a-d, the local field fluctuations are even larger than those at $\omega = \omega_{\mathrm{r}}$. At long wavelengths, clusters of conducting particles (rather than individual particles) resonate with the external field. Therefore, it is not surprising that the local field distributions are quite different from those in Fig. 5.2.

Figure 5.3 shows the field distributions at the percolation threshold $p = p_{\mathrm{c}} = 0.5$ for different wavelengths, namely $\lambda = 0.5\,\mu\mathrm{m}$ (Fig.5.3a), $\lambda = 1.5\,\mu\mathrm{m}$ (Fig. 5.3b), $\lambda = 10\,\mu\mathrm{m}$ (Fig. 5.3c), and $\lambda = 20\,\mu\mathrm{m}$ (Fig. 5.3d). Note that the field intensities in peaks increase with λ, and saturate at very high values

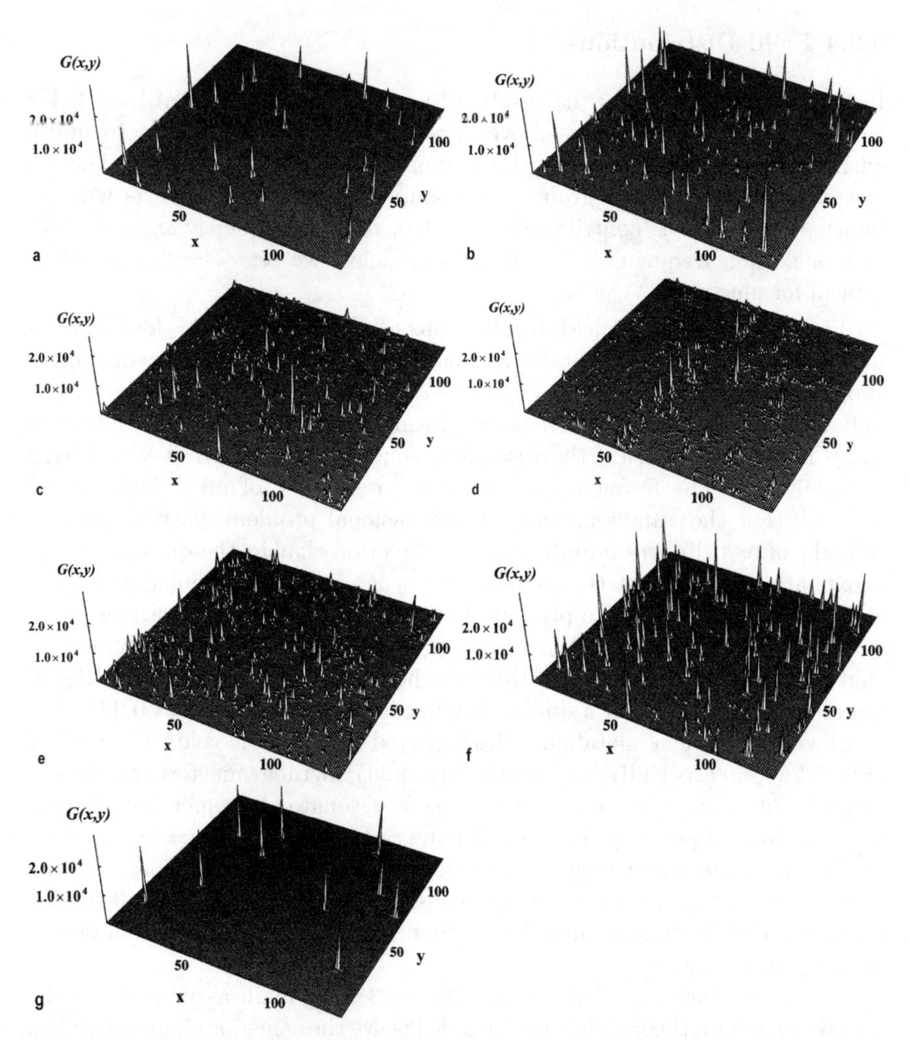

Fig. 5.2. Distribution of the local field intensities on a metal (silver) semicontinuous film for $\varepsilon'_{\rm m} = -\epsilon_{\rm d} = -2.2$ ($\lambda \approx 365$ nm) at different metal concentrations p. (**a**) $p = 0.001$; (**b**) $p = 0.01$; (**c**) $p = 0.1$; (**d**) $p = 0.5$; (**e**) $p = 0.9$; (**f**) $p = 0.99$; and (**g**)$p = 0.999$

$\sim 10^5|E^{(0)}|^2$; the peak spatial separations increase with λ. In the next section, we consider a scaling theory for the field distributions on a semicontinuous film that explains the above simulation results.

We also note that spatial locations of the field peaks strongly depend on the frequency. Qualitatively similar results were previously demonstrated for fractals and self-affine films considered in Chaps. 3 and 4. By changing the frequency one can excite different nm-size "hot spots" on a film. This effect is of high potential for various applications and can be studied exper-

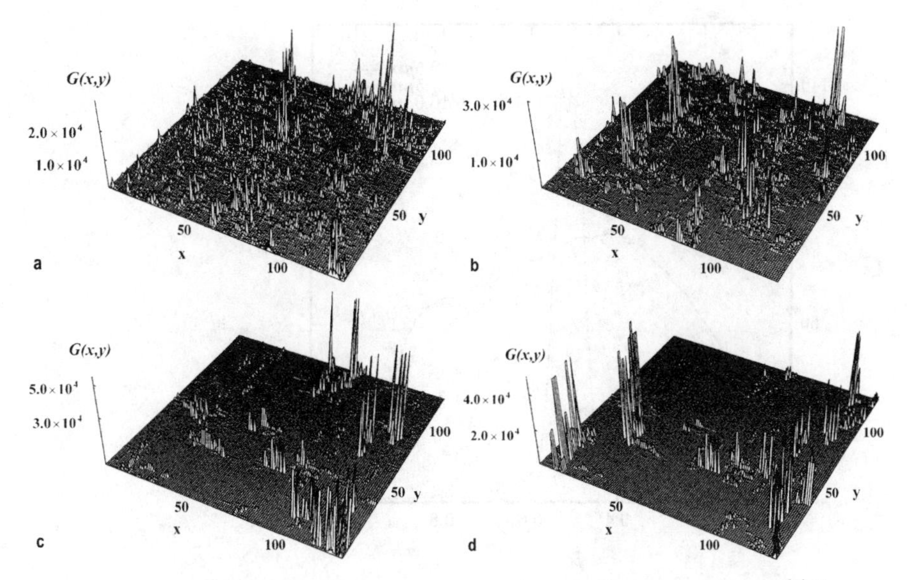

Fig. 5.3. Distribution of the local field intensities $G(x, y) = |E(\mathbf{r})/E^{(0)}|^2$ on a semicontinuous film at the percolation threshold for different wavelengths; (**a**) $\lambda = 0.5\,\mu\text{m}$; (**b**) $\lambda = 1.5\,\mu\text{m}$; (**c**) $\lambda = 10\,\mu\text{m}$; and (**d**) $\lambda = 20\,\mu\text{m}$

imentally using near-field scanning optical microscopy (NSOM) providing a sub-wavelength resolution [47, 77, 124].

Figure 5.4 shows the average enhancements of the intensity of the local fields $G = \left\langle |\mathbf{E}(\mathbf{r})|^2 \right\rangle / \left|\mathbf{E}^{(0)}\right|^2$. The results are shown as a function of p for different wavelengths $\lambda = 0.5\,\mu\text{m}$, $\lambda = 1.5\,\mu\text{m}$, and $\lambda = 10\,\mu\text{m}$. We see that field enhancements are large on average (a factor of $\sim 10^2$), but much smaller than in the local peaks in Fig. 5.3. This is because the sharp peaks are separated by relatively large distances so that the average enhancement is not as large as the local one in the peaks. The other high-order moments of the field distribution, which are important for estimation of the nonlinear response, experience even stronger enhancement and will be considered below.

The range of values of p where the enhancement occurs is large in the optical spectral range. However, it shrinks toward the larger wavelengths, as seen in Fig. 5.4.

From the above results, it follows that the local fields experience strong space fluctuations on a semicontinuous film; the large fields in the peaks result in giant enhancements of the optical nonlinearities considered below.

Figure 5.5 shows the distribution for the local enhancements of second-harmonic generation (SHG) at $p = p_c$, $g_{\text{SHG}}(\mathbf{x}, \mathbf{y}) = |E(x, y)/E^{(0)}|^4$, for different wavelengths, $\lambda = 0.36\,\mu\text{m}$, $\lambda = 0.5\,\mu\text{m}$, $\lambda = 1.5\,\mu\text{m}$, $\lambda = 10\,\mu\text{m}$ and $\lambda = 20\,\mu\text{m}$ (we assume that there is no additional enhancement at the

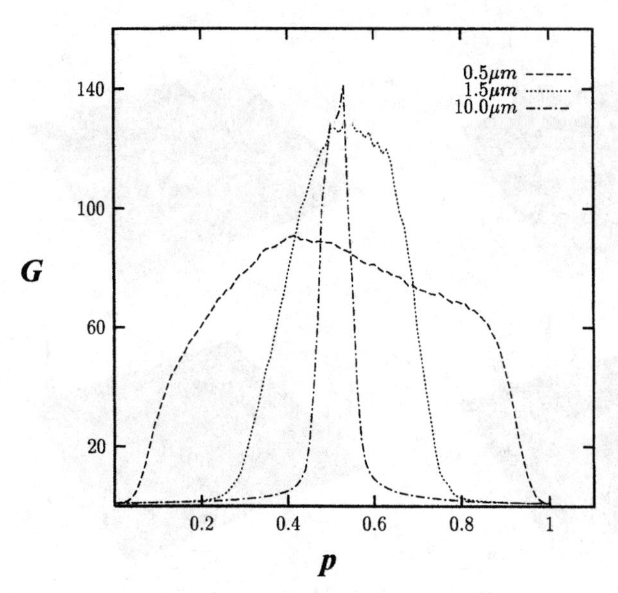

Fig. 5.4. Average enhancement of the field intensity, $G = \left\langle |\mathbf{E}(\mathbf{r})|^2 \right\rangle / \left|\mathbf{E}^{(0)}\right|^2$, on a silver semicontinuous film as a function of the metal concentration p for three different wavelengths

generated frequency). Qualitatively similar distribution also occurs for local enhancement of Raman scattering (RS).

As seen in the figure, the local enhancements have the form of sharp peaks sparsely distributed on the film, with the magnitudes increasing toward the long-wavelength part of the spectrum and saturating at very high values $\sim 10^{12}$. The average enhancement for both SHG and for RS as shown below, is much less (for G_{RS}, for example, we have $G_{\mathrm{RS}} = \left\langle |E(x,y)/E^{(0)}|^4 \right\rangle \sim 10^6$). According to Fig. 5.5, the peak positions strongly depend on the frequency. This nontrivial pattern for the local enhancement distribution can be probed by means of near-field optical microscopy [124]. If the density of SHG- and/or Raman-active molecules is small enough, each peak can be due to SHG and/or RS from *single* molecules. Thus the picture presented makes feasible Raman and nonlinear spectroscopy of single molecules on a semicontinuous metal film.

The above distributions for local enhancements shown in Fig. 5.5 result from strong fluctuations of the local fields (and currents) near the percolation threshold. Since the enhancement is proportional to the local field raised to the fourth power the largest contribution to the signal comes from small areas where the fluctuating fields are especially high. These areas of large local fields are associated with the film eigenmodes, which resonate at a given laser wavelength λ. When λ is changed, new eigenmodes of a film are excited,

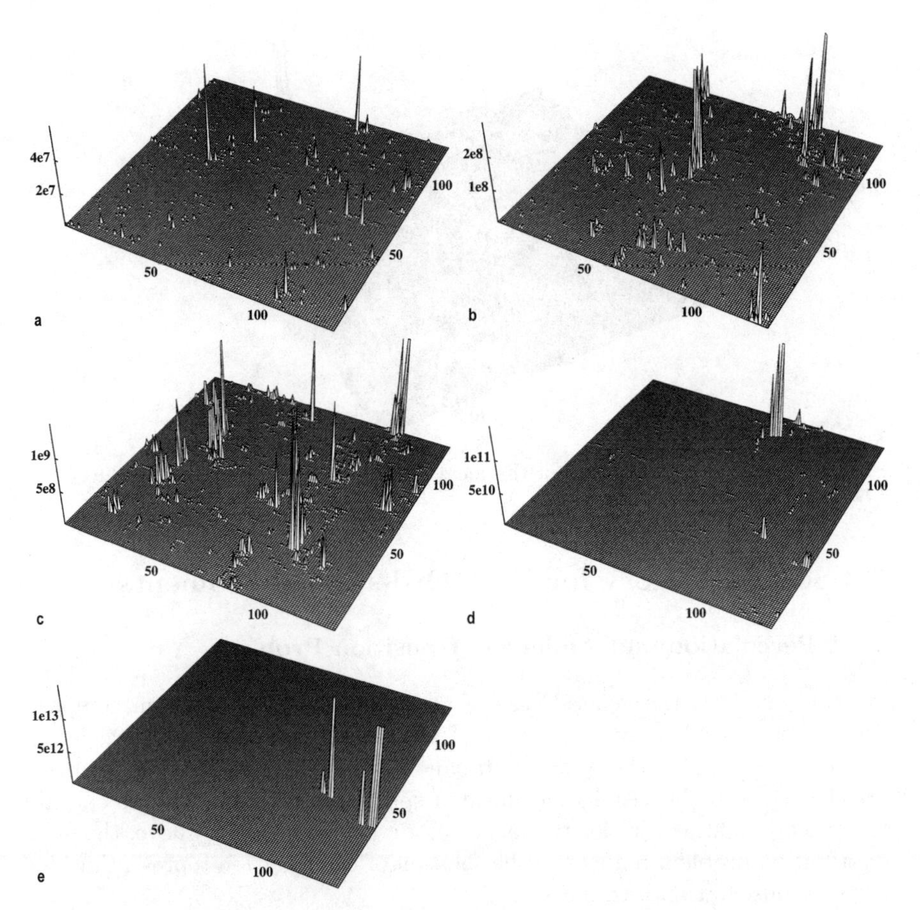

Fig. 5.5. Distributions of the local SHG enhancements $g_{\mathrm{SHG}}(\mathbf{r})$ on a silver semicontinuous film at the percolation threshold for different wavelengths: (**a**) $\lambda = 0.36\,\mu\mathrm{m}$; (**b**) $\lambda = 0.5\,\mu\mathrm{m}$; (**c**) $\lambda = 1.5\,\mu\mathrm{m}$; (**d**) $\lambda = 10\,\mu\mathrm{m}$; and (**e**) $\lambda = 20\,\mu\mathrm{m}$

resulting again in high local fields in the areas where these new modes are located.

Figure 5.6 shows the distribution $g_{\mathrm{THG}}(\mathbf{r})$ [see (5.33)] for the surface-enhanced local THG signals at $\lambda = 1.5\,\mu\mathrm{m}$. We can see that, similar to Fig. 5.5, the local THG signals also consist of spatially separated sharp peaks, as expected. The local enhancements can be huge: up to a factor of 10^{13} for $\lambda \sim 1\,\mu\mathrm{m}$, and 10^{16} for $\lambda \sim 10\,\mu\mathrm{m}$ (not shown). We emphasize again that enhancement in the peaks significantly exceeds the surface-average enhancement. The reason for this is in part the destructive interference between fields in different points, and in part the fact that the peaks are separated by distances significantly larger than their spatial sizes.

Next we consider a scaling theory that explains the described distributions of the local fields at the fundamental and generated frequencies.

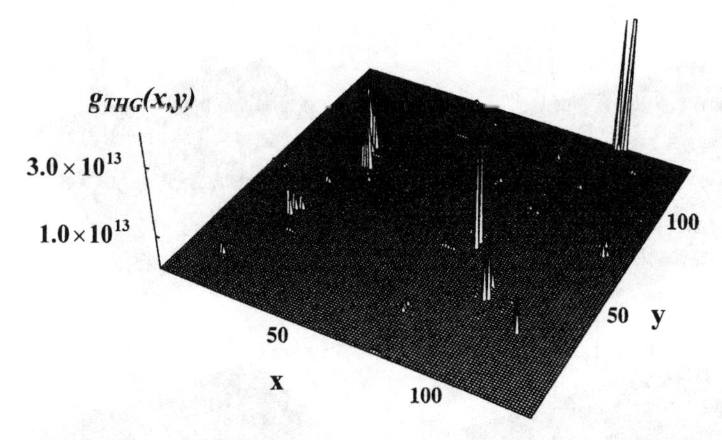

Fig. 5.6. Spatial distributions of the local THG enhancements, $g_{\mathrm{THG}}(\mathbf{r})$, for $\lambda = 1.5\,\mu\mathrm{m}$ $(p = p_{\mathrm{c}})$

5.3 Scaling Theory for High-Order Field Moments

5.3.1 Percolation and Anderson Transition Problem

We consider here the general case of a three-dimensional random composite. As mentioned, a typical size a of the metal grains in the percolation nanocomposites is of the order of 10 nanometers, i.e. much smaller then the wavelength λ in the visible and infrared spectral ranges, so that we can introduce a potential $\phi(\mathbf{r})$ for the local electric field. As shown above, the field distribution problem reduces to the solution of the equation representing the current conservation law, namely

$$\nabla \cdot \left(\sigma(\mathbf{r})\left[-\nabla\phi(\mathbf{r}) + \mathbf{E}^{(0)}(\mathbf{r})\right]\right) = 0, \tag{5.34}$$

where $\mathbf{E}^{(0)}$ is the applied field (we set $\mathbf{E}_{\mathrm{e}}(\mathbf{r}) = \mathbf{E}^{(0)}(\mathbf{r})$) and $\sigma(\mathbf{r})$ is the local conductivity that takes σ_{m} and σ_{d} values, for the metal and dielectric, respectively. In the discretized form, this relation acquires the form of Kirchhoff's equations defined, for example, on a cubic lattice [20].

We can write Kirchhoff's equations in terms of the local dielectric constant $\epsilon = (4\pi\mathrm{i}/\omega)\sigma$, rather than conductivity σ. We assume that the external electric field $\mathbf{E}^{(0)}$ is directed, say, along the x axis. Thus, in the discretized form (5.34), is equivalent to the following set of equations:

$$\sum_j \epsilon_{ij}\left(\phi_i - \phi_j + E_{ij}\right) = 0, \tag{5.35}$$

where ϕ_i and ϕ_j are the electric potentials determined at the sites of the cubic lattice (or the square lattice, for a two-dimensional system); the summation is over the six nearest neighbors of the site i. The electromotive force E_{ij} takes the value $E^{(0)}a_0$ for the bond $\langle ij\rangle$ aligned in the positive x direction, where

a_0 is the spatial period of the cubic lattice (which can coincide with the size grain a). For the bond $\langle ij \rangle$ aligned in the $-x$ direction, the electromotive force E_{ij} takes the value $-E^{(0)}a_0$; for the other four bonds related to site i, we have $E_{kj} = 0$. The permittivities ϵ_{ij} take values ϵ_m and ϵ_d, with probabilities p and $1-p$ respectively. Thus a percolation composite is modeled by a random network, including the electromotive forces E_{ij} that represent the external field.

For simplicity, we can assume that the cubic lattice has a very large but finite number of sites N and rewrite (5.35) in the matrix form with the "interaction matrix" $\widehat{\mathrm{H}}$ defined in terms of the local dielectric constants:

$$\widehat{\mathrm{H}}\,\phi = \mathcal{E}, \tag{5.36}$$

where ϕ is the vector of the local potentials $\phi = \{\phi_1, \phi_2, \ldots, \phi_N\}$ determined at N sites of the lattice. Vector $\mathcal{E}$ also has N components $\mathcal{E}_i = \sum_j \epsilon_{ij} E_{ij}$, as follows from (5.35). The $N \times N$ matrix $\widehat{\mathrm{H}}$ has off-diagonal elements $H_{ij} = -\epsilon_{ij}$ and diagonal elements $H_{ii} = \sum_j \epsilon_{ij}$ (j refers to the nearest neighbors of the site i). The off-diagonal elements H_{ij} take randomly values $\epsilon_{\mathrm{d}} > 0$ and $\epsilon_{\mathrm{m}} = (-1 + \mathrm{i}\kappa)\,|\epsilon'_{\mathrm{m}}|$, where the loss-factor $\kappa = \epsilon''_{\mathrm{m}}/\,|\epsilon'_{\mathrm{m}}|$ is small in the optical and infrared spectral ranges, i.e. $\kappa \ll 1$. The diagonal elements H_{ii} are also random numbers distributed between $2d\epsilon_{\mathrm{m}}$ and $2d\epsilon_{\mathrm{d}}$, where $2d$ is the number of nearest neighbors in the lattice for a d-dimensional system.

It is important to note that the matrix $\widehat{\mathrm{H}}$ is similar to the quantum-mechanical Hamiltonian for Anderson's transition problem with both on- and off-diagonal correlated disorder [157, 158]. We will refer hereafter to operator $\widehat{\mathrm{H}}$ as to Kirchhoff's Hamiltonian (KH). In the approach considered here, the field distribution problem, i.e. the problem of finding a solution to the system of linear equations (5.35), can be translated into the problem of finding the eigenmodes of KH $\widehat{\mathrm{H}}$.

Suppose we have found the eigenvalues Λ_n for $\widehat{\mathrm{H}}$. In the optical and infrared spectral ranges, the real part ϵ'_{m} of the metal dielectric function ϵ_{m} is negative ($\epsilon'_{\mathrm{m}} < 0$), whereas the permittivity of a dielectric host is positive ($\epsilon_{\mathrm{d}} > 0$); as mentioned, the lossfactor is small, and so $\kappa = \epsilon''_{\mathrm{m}}/\,|\epsilon'_{\mathrm{m}}| \ll 1$. Therefore, the manifold of the KH eigenvalues Λ_n contains the eigenvalues that have their real parts equal (or close) to zero with very small imaginary parts ($\kappa \ll 1$). Then the eigenstates that correspond to the eigenvalues $|\Lambda_n/\epsilon_{\mathrm{m}}| \ll 1$ are strongly excited by the external field and are seen as giant field fluctuations, representing nonuniform plasmon resonances of a percolation system.

The localized optical excitations can be thought of as field peaks separated on average by the distance $\xi_{\mathrm{e}} \propto a\,(N/n)^{1/d}$, where n is the number of the resonant KH eigenmodes excited by the external field and N is the total number of the eigenstates. In the limit $\kappa \ll 1$, only a small part $n \sim \kappa N$ of the eigenstates is effectively excited by the external field. Therefore, the distance ξ_{e}, which we call the field correlation length, is large: $\xi_{\mathrm{e}}/a \propto \kappa^{-1/d} \gg 1$.

5.3.2 Resonance Excitations at $\epsilon_{\mathrm{d}} = -\epsilon'_{\mathrm{m}}$

To estimate the giant field fluctuations, we first consider the important case $\omega \simeq \omega_{\mathrm{r}}$, where ω_{r} corresponds to the resonance of individual particles occurring (for $d = 2$) at $\epsilon_{\mathrm{d}} = -\epsilon'_{\mathrm{m}}$. For further analysis it is convenient to normalize the permittivities as $\epsilon_m = -1 + i\kappa$ and $\epsilon_d = 1$.

The Kirchhoff equations (5.36) can be written then as $(H' + i\kappa H'')\,\phi =$, where H' and H'' are both Hermitian, and $\kappa H'' \ll 1$ represents small losses in the system. Then we can express the potential Φ in terms of the eigenfunctions Ψ_n of H' (see below).

According to the one-parameter scaling theory, the eigenstates Ψ_n are thought to be all localized for the two-dimesional case (see, however, the discussion in [159, 160]). On the other hand, it was shown that there is a transition from chaotic eigenstates [161, 162] to localized eigenstates in the two-dimensinal Anderson problem [163] with intermediate crossover region. We consider first the case when the metal concentration p equals to the percolation threshold $p = p_{\mathrm{c}} = 1/2$, for the two-dimensional bond percolating problem. Then, the on-diagonal disorder in the KH is characterized by $\langle \mathrm{H}'_{ii} \rangle = 0$, $\left\langle \mathrm{H}'^2_{ii} \right\rangle = 4$, which corresponds to the chaos-localization transition [163]. The KH also has strong off-diagonal disorder, $\left\langle \mathrm{H}'_{ij} \right\rangle = 0$ $(i \neq j)$, which usually favors localization [164, 165]. Our conjecture is that the eigenstates Ψ_n are localized, at least those with $\Lambda_n \approx 0$, in a two-dimensional system. (We cannot, however, rule out a possibility of inhomogeneous localization similar to that obtained for fractals [48, 49] or power-law localization [157, 166].)

In the case where $\epsilon_{\mathrm{d}} = -\epsilon'_{\mathrm{m}} = 1$ and $p = 1/2$, all parameters in KH $\widehat{\mathrm{H}}'$ are of the order of one and its properties do not change under the transformation $\epsilon_{\mathrm{d}} \Longleftrightarrow \epsilon_{\mathrm{m}}$. Therefore the real eigenvalues Λ_n are distributed symmetrically with respect to zero. The resonating eigenstates with $\Lambda_n \approx 0$ are most effectively excited by the applied field. When the metal concentration p decreases (increases) the eigenstates with $\Lambda_n \approx 0$ shift from the center of the distribution toward its tails, which typically favors localization so that localization is preserved and can be even stronger for $p \neq p_{\mathrm{c}}$.

The Anderson transition in a three-dimensional system is less understood and little is known about the eigenfunctions [157, 167]. Below, we conjecture that the eigenstates with $\Lambda_n \approx 0$ are also localized in the three-dimensional case.

We expand now the fluctuating potential ϕ over the eigenstates Ψ_n (with eigenvalues Λ_n) of the real (Hermitian) part H' of the KH: $\mathrm{H} = \mathrm{H}' + \mathrm{i}\mathrm{H}''$. By substituting $\phi = \sum_n A_n \Psi_n$ in (5.36), we obtain

$$(\mathrm{i}\kappa b_n + \Lambda_n)\, A_n + \mathrm{i}\kappa \sum_{m \neq n} (\Psi_n \mid H'' \mid \Psi_m)\, A_m = \mathcal{E}_n, \tag{5.37}$$

where $b_n = \left(\Psi_n \mid \widehat{\mathrm{H}}'' \mid \Psi_n\right)$ and $\mathcal{E}_n = (\Psi_n \mid \mathcal{E})$ is the projection of the external field on the eigenstate Ψ_n. We suggest that the eigenstates Ψ_n are localized in spatial domains $\xi_{\mathrm{A}}(\Lambda)$, where $\xi_{\mathrm{A}}(\Lambda)$ is the Anderson correlation length, which depends on eigenvalue Λ. (From Fig. 5.2d we roughly estimate that ξ_{A} is between a and $10a$.) Then the sum in (5.37) converges and may be treated as a small perturbation. In the first approximation $A_n = \mathcal{E}_n/(\Lambda_n + \mathrm{i}\kappa b_n)$. A small correction to A_n is given by $A_n' = -\mathrm{i}\kappa \sum_{m\neq n} \left(\Psi_n \mid \widehat{\mathrm{H}}'' \mid \Psi_m\right) \mathcal{E}_m/(\Lambda_m + \mathrm{i}\kappa b_m)$. For $\kappa \to 0$, most important modes with $|\Lambda_m| \leq \kappa$ have the spatial density $a^{-d}\kappa \to 0$. Therefore A_n' is exponentially small and can be neglected. Since all parameters in the real Hamiltonian $\widehat{\mathrm{H}}''$ are of the order of one, the matrix elements b_n are also of the order of one. For the eigenvalues of interest, $|\Lambda_m| \leq \kappa$, we approximate them by some constant b, which is about unity. Then, the local potential ϕ is given by $\phi(\mathbf{r}) = \sum_n \mathcal{E}_n \Psi_n(r)/(\Lambda_n + \mathrm{i}\kappa b)$ and the fluctuating part of the local field $\mathbf{E}_{\mathrm{f}} = -\nabla\phi(\mathbf{r})$ is given by

$$\mathbf{E}_f(\mathbf{r}) = -\sum_n \mathcal{E}_n \nabla\Psi_n(\mathbf{r})/(\Lambda_n + \mathrm{i}\kappa b)\,. \tag{5.38}$$

The average field intensity is given by

$$\left\langle |E|^2 \right\rangle = \left\langle |\mathbf{E}_{\mathrm{f}} + \mathbf{E}_0|^2 \right\rangle = E_0^2 + \left\langle \sum_{n,m} \frac{\mathcal{E}_n \mathcal{E}_m^* \left(\nabla\Psi_n(\mathbf{r}) \cdot \nabla\Psi_m^*(\mathbf{r})\right)}{(\Lambda_n + \mathrm{i}\kappa b)(\Lambda_m - \mathrm{i}\kappa b)} \right\rangle, \tag{5.39}$$

where we have taken into account that the average $\langle \mathbf{E}_f \rangle = 0$. (Note that for the applied field we use interchangeably the notations $\mathbf{E}^{(0)}$ and $\mathbf{E}_0$.) A localized eigenstate Ψ_n can be characterized by the "center" of localization $\mathbf{r}_n$, so that $\Psi_n = \Psi(\mathbf{r} - \mathbf{r}_n)$. Now we consider the eigenstates $\Psi(\mathbf{r} - \mathbf{r}_n)$ with eigenvalues Λ_n within a small interval $|\Lambda_n - \Lambda| \ll \kappa$ centered at Λ. For all realizations of a macroscopically homogeneous random film, positions $\mathbf{r}_n$ of eigenfunctions $\Psi(\mathbf{r} - \mathbf{r}_n)$ take different values that do not correlate with the value of Λ. Therefore we can independently average the numerator in (5.39) over positions $\mathbf{r}_n$ and $\mathbf{r}_m$ of the eigenstates Ψ_n and Ψ_m corresponding to eigenvalues Λ_n and Λ_m, respectively. Taking into account that $\langle \nabla\Psi_n(\mathbf{r}) \rangle = 0$, we obtain

$$\left\langle |E|^2 \right\rangle = E_0^2 + \sum_n \frac{|\mathcal{E}_n|^2 \left\langle |\nabla\Psi_n(\mathbf{r})|^2 \right\rangle}{\Lambda_n^2 + (b\kappa)^2}. \tag{5.40}$$

Localized eigenstates are not degenerate in general so that the eigenfunctions Ψ_n can be chosen as real. Then we can estimate $|\mathcal{E}_n|$ in (5.40) as $|\mathcal{E}_n|^2 = |(\Psi_n \mid \mathcal{E})|^2 \sim a^{4-2d} \left|\int \Psi_n (\mathbf{E}_0 \cdot \nabla\epsilon)\,\mathrm{d}\mathbf{r}\right|^2 = a^{4-2d} \left|\int \epsilon (\mathbf{E}_0 \cdot \nabla\Psi_n)\,\mathrm{d}\mathbf{r}\right|^2 \sim E_0^2 a^2 (\xi_{\mathrm{A}}(\Lambda_n)/a)^{d-2}$, and $\left\langle |\nabla\Psi_n(\mathbf{r})|^2 \right\rangle \sim \xi_{\mathrm{A}}^{-2}(\Lambda_n)/N$. Substitution of these estimates in (5.40) gives

$$\left\langle |E|^2 \right\rangle \sim E_0^2 \int \frac{\rho(\Lambda)\,(a/\xi_{\mathrm{A}}(\Lambda))^{4-d}}{\Lambda^2 + (b\kappa)^2} \mathrm{d}\Lambda, \tag{5.41}$$

where $\rho(\Lambda) = a^d \sum_n \delta(\Lambda - \Lambda_n)/V$ is the dimensionless density of states for the KH $\widehat{\mathrm{H}}'$ and V is the system volume. We have taken into account that the second (fluctuating) term in (5.40) can be much larger than the first one. Since all matrix elements in H' are of the order of one, the second moment of the local electric field $M_2 \equiv M_{2,0} = \left\langle |E|^2 \right\rangle / E_0^2$ is [for the general definition see (5.45)] estimated as

$$M_2^* \sim \rho\,(a/\xi_\mathrm{A})^{4-d}\,\kappa^{-1} \gg 1, \tag{5.42}$$

provided that $\kappa \ll \rho\,(a/\xi_\mathrm{A})^{4-d}$ (we set $\xi_\mathrm{A}(0) \equiv \xi_\mathrm{A},\ \rho(0) \equiv \rho$ and used $b \sim 1$). Thus the field distribution can in this case be described as a set of the KH eigenstates localized within ξ_A, with the field peaks having the amplitudes

$$E_\mathrm{m}^* \sim E_0 \kappa^{-1} (a/\xi_\mathrm{A})^2, \tag{5.43}$$

which are separated in distance by the field correlation length

$$\xi_\mathrm{e}^* \sim a(\rho\kappa)^{1/d} \gg \xi_\mathrm{A}. \tag{5.44}$$

Hereafter, by the superscript * we mark the quantities whose values must be taken at the condition $-\epsilon_\mathrm{m}' = \epsilon_\mathrm{d} = 1$ considered here (this sign, of course, should not be confused with complex conjugation). Using the scale renormalization described in the next section, we will see how these quantities are transformed when $|\epsilon_\mathrm{m}/\epsilon_\mathrm{d}| \gg 1$, i.e. in the long wavelength part of the spectrum. (Note however that, for ξ_A and ρ, we omit the * sign in order to avoid complicated notations; it is implied that their values are always taken at $-\epsilon_\mathrm{m}' = \epsilon_\mathrm{d} = 1$, even if other qunatities are taken at different values of the dielectric permittivities .)

The above results are in good agreement with comprehensive numerical calculations performed in [148]–[150] for a two-dimensional system with $\epsilon_\mathrm{m}/\epsilon_\mathrm{d} \approx -1$ and $p = p_\mathrm{c}$. It was shown in [148]–[150] that the average intensity of the local field fluctuations $\langle |E(\mathbf{r})|^2 \rangle$ is estimated as $\langle |E(\mathbf{r})|^2 \rangle / \left|\mathbf{E}^{(0)}\right|^2 = M_2^* \sim \kappa^{-\gamma}$, with the critical exponent $\gamma \approx 1$. The authors also found that the correlation length ξ_e^* of the field fluctuations scales as $\xi_\mathrm{e}^* \sim \kappa^{-\nu_\mathrm{e}}$, with the critical exponent $\nu_\mathrm{e} \approx 0.5$, for $d = 2$. These values of the critical exponents γ and ν_e are very close to those found above.

Now we consider arbitrary high-order field moments defined as

$$M_{n,m} = \frac{1}{V E_0^m |E_0|^n} \int |E(\mathbf{r})|^n E^m(\mathbf{r})\, d\mathbf{r} \tag{5.45}$$

where, as above, $E_0 \equiv E^{(0)}$ is the amplitude of the external field and $E(\mathbf{r})$ (note that $E^2(\mathbf{r}) \equiv \mathbf{E}(\mathbf{r}) \cdot \mathbf{E}(\mathbf{r})$) is the local field; the integration is over the total volume V of the system.

The high-order field moment $M_{2k,m} \propto E^{k+m} E^{*k}$ represents a nonlinear optical process in which, in one elementary act, $k+m$ photons are added and k photons are subtracted [55]. This is because the complex conjugated field in the general expression for the nonlinear polarization implies photon subtraction, so that the corresponding frequency enters the nonlinear susceptibility with the minus sign [55]. Enhancement of the Kerr optical nonlinearity G_{K} is equal to $M_{2,2}$, SHG and THG enhancements are given by $|M_{0,2}|^2$ and $|M_{0,3}|^2$, and SERS is represented by $M_{4,0}$.

We are interested here in the case when $M_{n,m} \gg 1$, which implies that the fluctuating part of the local electric field $\mathbf{E}_{\mathrm{f}}$ is much larger than the applied field $\mathbf{E}_0$. We substitute in (5.45) the expression for $\mathbf{E}_{\mathrm{f}}$ given by (5.38) and average over spatial positions of the eigenstates $\Psi_n(\mathbf{r}) \equiv \Psi(\mathbf{r}-\mathbf{r}_n)$; this results in the following estimate:

$$M_{n,m} \sim \int \frac{\rho(\Lambda)\left(a/\xi_{\mathrm{A}}(\Lambda)\right)^{2(n+m)-d}}{\left(\Lambda^2+(b\kappa)^2\right)^{n/2}\left(\Lambda+ib\kappa\right)^m}\mathrm{d}\Lambda. \tag{5.46}$$

Note that to obtain the above expression we neglect all cross-terms in the product of eigenstates when averaging over their spatial positions. It can be shown that, after integrating over Λ, these cross-terms result in negligible [in comparison with the leading term given by (5.46)] contribution to $M_{n,m}$ for $\kappa \to 0$.

Assuming that the density of states $\rho(\Lambda)$ and the localization length $\xi_{\mathrm{A}}(\Lambda)$ are both smooth functions of Λ in the vicinity of zero, and taking into account that all parameters of $\widehat{\mathrm{H}}'$ for the case $-\epsilon'_{\mathrm{m}} = \epsilon_{\mathrm{d}} = 1$ are of the order of unity, we obtain the following estimate for the local field moments:

$$M^*_{n,m} \sim \rho(p)\left(a/\xi_{\mathrm{A}}(p)\right)^{2(n+m)-d}\kappa^{-n-m+1}, \tag{5.47}$$

for $n+m>1$ and $m>0$. Note that the same estimate can be obtained by considering the local fields as a set of peaks (stretched out for the distance ξ_{A}), with magnitude E^*_{m} and average distance ξ^*_{e} between the peaks. Recall that the superscript * marks physical quantities defined in the system $-\epsilon'_{\mathrm{m}} = \epsilon_{\mathrm{d}} = 1$, and this sign is omitted for ρ and ξ_{A}, which are always given at the condition $-\epsilon'_{\mathrm{m}} = \epsilon_{\mathrm{d}} = 1$.

In (5.47) we indicated explicitly dependence of the density of states $\rho(p)$ and localization length $\xi_{\mathrm{A}}(p)$ on the metal concentration p. The notations $\rho(p)$ and $\xi_{\mathrm{A}}(p)$ should be understood as the density of states $\rho(p,\Lambda=0)$ and localization length $\xi_{\mathrm{A}}(p,\Lambda=0)$, i.e. they are given at $\Lambda=0$.

It is clear that for a small metal concentration p, the eigenvalue distribution shifts to the positive side, so that the eigenstates with $\Lambda \approx 0$ shift to the lower edge of the spectrum. The number of states excited by the external field is proportional to the number of metal particles, when $p \to 0$, i.e. $\rho(p) \sim p$, for $p \to 0$. The same consideration holds in another limit, when a small number of holes in an otherwise continuous metal film resonate with the external field, i.e. $\rho(p) \sim 1-p$ for $p \to 1$. We anticipate that the density of states $\rho(p)$ has a maximum when the eigenvalue $\Lambda = 0$ coincides with the center of the spectrum of the eigenvalues of the KH $\widehat{\mathrm{H}}'$, i.e. at $p = p_{\mathrm{c}} = 1/2$, in the case of $d=2$.

The Anderson localization length $\xi_{\mathrm{A}}(\Lambda)$ has, typically, its maximum at the center of the Λ distribution [157]. When p diverges from p_{c}, the value $\Lambda = 0$ moves from the center of the Λ-distribution toward its wings, where the localization is typically stronger (i.e. ξ_{A} is less). Therefore, it is plausible to suggest that $\xi_{\mathrm{A}}(p)$ reaches its maximum at $p = p_{\mathrm{c}} = 1/2$ and decreases toward $p = 0$ and $p = 1$, so that the local field moments may have a minimum at $p = p_{\mathrm{c}} = 1/2$, in accordance with (5.47).

It is important to note that the moment magnitudes in (5.47) do not depend on the number of "subtracted" (annihilated) photons, in one elementary act of the nonlinear scattering. If there is at least one such photon, then the poles in (5.46) are in different complex semiplanes and the result of the integration is estimated by (5.47).

However, for the case when all photons are added (in other words, all frequencies enter the nonlinear susceptibility with positive sign); i.e. when $n = 0$, we cannot estimate the moments $M_{0,m} \equiv (\mathbf{E}^{(0)})^{-m} V^{-1} \int E^m(\mathbf{r})\, d\mathbf{r}$ by (5.47) since the integral in (5.46) is not further determined by the poles at $\Lambda = \pm i b \kappa$. Yet all the functions of the integrand are about unity and the moment $M_{0,m}$ must be of the order of unity $M_{0,m} \sim \mathrm{O}(1)$ for $m > 1$. Note that the moment $M_{0,m}$ describes, in particular, the enhancement of n-order harmonic generation, through the relation $G_{n\mathrm{HG}} = |M_{0,m}|^2$.

5.3.3 Scaling Renormalization and Collective Resonances at $|\epsilon_{\mathrm{m}}|/\epsilon_{\mathrm{d}} \gg 1$

Above we assumed that $|\epsilon_{\mathrm{m}}'|/\epsilon_{\mathrm{d}} = 1$, which for two-dimensional systems corresponds to the resonance condition of individual particles. To find the local field distributions in a percolation composite for a large contrast, $\mathcal{H} = |\epsilon_{\mathrm{m}}|/\epsilon_{\mathrm{d}} \gg 1$, we further develop the scaling arguments of the percolation theory [13, 149, 150]. In the case of the high contrast, the resonance optical excitations cannot occur in individual particles; instead, the resonance is due to collective excitations of clusters of particles.

We divide a system into cubes (or squares, for the two-dimensional case) of size l, with l much less than the percolation length, i.e. $l \ll \xi$, and we consider each cube as a new renormalized element. All such cubes can be classified into two types: a cube that contains a continuous path of metallic particles is considered a "conducting" element; and a cube without such an "infinite" cluster is considered a nonconducting, "dielectric", element. Following the finite-size arguments [see (1.26) and (1.27)], the effective dielectric constant of the "conducting" cube $\epsilon_{\mathrm{m}}(l)$ decreases with increasing size l as $\epsilon_{\mathrm{m}}(l) \cong l^{-t/\nu}\epsilon_{\mathrm{m}}$, whereas the effective dielectric constant of the "dielectric" cube $\epsilon_{\mathrm{d}}(l)$ increases with l as $\epsilon_{\mathrm{d}}(l) \cong l^{s/\nu}\epsilon_{\mathrm{d}}$ (t and s are the percolation-critical exponents for the static conductivity and dielectric constant respectively, and ν is the critical index for the percolation correlation length: $\xi \cong |p - p_{\mathrm{c}}|^{-\nu}$ [see (1.28)]. For the two-dimensional case, $t \cong s \cong \nu \cong 4/3$; for three dimensions, the exponents are given by $t = 2.05$, $s = 0.76$, and $\nu = 0.89$ [20]).

We set now the cube size l to be equal to l_r which is defined as

$$l_r = a(|\epsilon_m|/\epsilon_d)^{\nu/(t+s)}. \tag{5.48}$$

Then, in the renormalized system, where each cube of the size l_r is considered as a single element, the ratio of the dielectric constants of these new elements is equal to $\epsilon_m(l_r)/\epsilon_d(l_r) \cong \epsilon_m/|\epsilon_m| \approx -1 + i\kappa$, where the lossfactor $\kappa \approx \omega_\tau/\omega \ll 1$. Thus, after the renormalization, the problem becomes equivalent to the field distribution considered above for the case $\epsilon_d \simeq -\epsilon'_m \sim 1$.

We note that according to the renormalization group transformation [13], the concentration of conducting and dielectric elements does not change under renormalization, provided that $p = p_c$.

To develop further insight into the problem, we consider now two neighboring conducting clusters of size l. The clusters have inductive conductances $\Sigma_i = -i\omega\epsilon_m(l)l/4\pi$; this is because metal conductivity is inductive for $\omega < \tilde{\omega}_p$ ($\epsilon'_m < 0$, $|\epsilon'_m| \gg \epsilon''_m$). The gap between the two conducting clusters has a capacitive conductance $\Sigma_c(l) \simeq -i\omega\epsilon_d(l)l/4\pi$. We choose the size $l = l_r$ so that the capacitive and inductive conductances are equal to each other in magnitude but opposite in sign, so that$|\Sigma_c(l_r)| = |\Sigma_i(l_r)|$. Then there is a resonance in the system, and the electric field is strongly enhanced in the intercluster gap, at the points of the closest approach where the gap shrinks down to a. At these points of the original system, the local field acquires the largest values.

Taking into account that the electric field renormalizes as $E^{(0)*} = E^{(0)}(l_r/a)$, we obtain that the field peaks are given by

$$\begin{aligned} E_m &\sim E^{(0)}\left(a/\xi_A\right)^2 (l_r/a)\kappa^{-1} \\ &\sim E^{(0)}(a/\xi_A)^2 \left(\frac{|\epsilon_m|}{\epsilon_d}\right)^{\nu/(t+s)} \left(\frac{|\epsilon_m|}{\epsilon''_m}\right), \end{aligned} \tag{5.49}$$

where $\xi_A \equiv \xi_A(p_c)$ is the localization length in the renormalized system.

The close-approach points determine the gap capacitive conductance $\Sigma_c(l_r)$, which in turn depends on the cluster size l_r. Therefore the number $n(l_r)$ of the close-approach points (or, in other words, the number of peaks in the resonance cube of the size l_r) scales with l_r in the same way as the conductance Σ_c, i.e. $n(l_r) \sim \Sigma_c(l_r) \sim \epsilon_d(l_r)l_r^{d-2} \sim l_r^{d-2+s/\nu}$.

The average distance between the field maxima in the renormalized system is equal to ξ_e^*; accordingly the average distance ξ_e between the chains of the field peaks in the original system (provided that $\rho \sim 1$) is

$$\xi_e \sim \xi_e^*(l_r/a) \sim a(|\epsilon_m|/\epsilon''_m)^{1/d}(|\epsilon_m|/\epsilon_d)^{\nu/(t+s)}. \tag{5.50}$$

The following pattern of the local field distribution emerges from the above consideration. The largest local fields of E_m amplitude are concentrated in a dielectric gap between the resonant metal clusters of size l_r. We can say that the field peaks of amplitude E_m^* from the localization "volume" ξ_A^d of the renormalized system are transformed into a chain of $(\xi_{\mathrm{A}}/a)^d n(l_{\mathrm{r}})$ peaks of the amplitude E_{m} in the original system. The "size" of the resonant state in the original system is $\xi_{\mathrm{A}}(l_{\mathrm{r}}/a)$ and spatial separations between these states are $\xi_{\mathrm{e}} \gg l_{\mathrm{r}}$.

With increasing wavelength, the scale l_{r} and the number of local field maxima $n(l_{\mathrm{r}})$ in each resonating cluster both increase. The average distance ξ_{e} between the chains of the peaks also increases with λ, so that ξ_{e} remains larger than l_{r}. The amplitudes of the field peaks $E_{\mathrm{m}}(\omega)$ increase with the wavelength until they reach the maximum value which for a Drude metal in a two-dimensional percolation system is estimated as $E_{\mathrm{m}}/E^{(0)} \sim \omega_{\mathrm{p}}/\omega_\tau \epsilon_{\mathrm{d}}^{1/2} \gg 1$. The described behavior for the field distributions can be tracked in Fig. 5.3.

Using the above scaling formulas, we obtain the following estimate for the local field moments $M_{n,m} \sim (E_{\mathrm{m}}/E^{(0)})^{n+m}(\xi_{\mathrm{A}}/a)^d n(l_{\mathrm{r}})/(\xi_{\mathrm{e}}/a)^d$, i.e.

$$\begin{aligned} M_{n,m} &\sim \left(\frac{E_{\mathrm{m}}}{E^{(0)}}\right)^{n+m} \frac{(l_{\mathrm{r}}/a)^{d-2+s/\nu}}{(\xi_{\mathrm{e}}/\xi_{\mathrm{A}})^d} \\ &\sim \rho(\xi_{\mathrm{A}}/a)^{d-2(n+m)} \left(\frac{l_{\mathrm{r}}/a}{\kappa}\right)^{n+m-1} (l_{\mathrm{r}}/a)^{s/\nu-1} \\ &\sim \rho(\xi_{\mathrm{A}}/a)^{d-2(n+m)} \left[\left(\frac{|\epsilon_{\mathrm{m}}|}{\epsilon_{\mathrm{d}}}\right)^{\nu/(t+s)}\right]^{n+m-2+s/\nu} \left(\frac{|\epsilon_{\mathrm{m}}|}{\epsilon''_{\mathrm{m}}}\right)^{n+m-1} \end{aligned} \tag{5.51}$$

for $n+m>1$ and $n>0$. Note that, since $s \approx \nu$ for both $d=2$ and $d=3$, the above formula can be simplified to

$$\begin{aligned} M_{n,m} &\sim \rho(\xi_{\mathrm{A}}/a)^{d-2(n+m)} \left(\frac{l_{\mathrm{r}}/a}{\kappa}\right)^{n+m-1} \\ &\sim \rho(\xi_{\mathrm{A}}/a)^{d-2(n+m)} \left[\left(\frac{|\epsilon_{\mathrm{m}}|}{\epsilon_{\mathrm{d}}}\right)^{\nu/(t+s)} \left(\frac{|\epsilon_{\mathrm{m}}|}{\epsilon''_{\mathrm{m}}}\right)\right]^{n+m-1} . \end{aligned} \tag{5.52}$$

Since $|\epsilon_{\mathrm{m}}| \gg \epsilon_{\mathrm{d}}$ and the ratio $|\epsilon_{\mathrm{m}}|/\epsilon''_{\mathrm{m}} \gg 1$, the moments of the local field are very large, i.e. $M_{n,m} \gg 1$, in the visible and infrared spectral ranges. Note that the first moment $M_{0,1} \simeq 1$, that corresponds to the equation $\langle \mathbf{E}(\mathbf{r}) \rangle = \mathbf{E}^{(0)}$.

Consider now the moments $M_{n,m}$ for $n=0$, i.e. $M_{0,m} = \langle E(\mathbf{r})^m \rangle / \left(E^{(0)}\right)^m$. In the renormalized system where $\epsilon_{\mathrm{m}}(l_{\mathrm{r}})/\epsilon_{\mathrm{d}}(l_{\mathrm{r}}) \cong -1 + i\kappa$, the field distribution coincides with the field distribution in the system with $\epsilon_{\mathrm{d}} \simeq -\epsilon'_{\mathrm{m}} \sim 1$. In that system, field peaks E^*_{m}, being different in phase, cancel each other out, resulting in the moment $M^*_{0,m} \sim \mathrm{O}(1)$ (see the discussion in the end of Subsect. 5.3.2). In transition to the original system, the peaks increase by the factor l_{r}, leading to an increase in the moment $M_{0,m}$. Then we obtain the following equation for the moment:

$$M_{0,m} \sim M_{0,m}^{*}(l_{\mathrm{r}}/a)^{m}\left(\frac{n(l_{\mathrm{r}})}{(\xi_{\mathrm{e}}/a)^{d}}\right) \sim \kappa(l_{\mathrm{r}}/a)^{m-2+s/\nu}$$

$$\sim \left(\frac{\epsilon_{\mathrm{m}}''}{|\epsilon_{\mathrm{m}}|}\right)\left(\frac{|\epsilon_{\mathrm{m}}|}{\epsilon_{\mathrm{d}}}\right)^{(m-2+s/\nu)\nu/(t+s)},$$

which holds when $M_{0,m}$ given by this equation is larger than unity. Since $s \approx \nu$ for both $d = 2$ and $d = 3$, the above formula can be simplified to

$$M_{0,m} \sim \kappa(l_{\mathrm{r}}/a)^{m-1} \sim \left(\frac{\epsilon_{\mathrm{m}}''}{|\epsilon_{\mathrm{m}}|}\right)\left(\frac{|\epsilon_{\mathrm{m}}|}{\epsilon_{\mathrm{d}}}\right)^{\frac{\nu(m-1)}{t+s}}. \tag{5.53}$$

For two-dimensional percolation composites, the critical exponents are given by $t \cong s \cong \nu \cong 4/3$, and (5.52) and (5.53) become (we assume that $\rho \sim 1$)

$$M_{n,m} \sim \left(\frac{|\epsilon_{\mathrm{m}}|^{3/2}}{(\xi_{\mathrm{A}}/a)^{2}\epsilon_{\mathrm{d}}^{1/2}\epsilon_{\mathrm{m}}''}\right)^{n+m-1} \quad (d=2), \tag{5.54}$$

for $n + m > 1$ and $n > 0$, and

$$M_{0,m} \sim \frac{\epsilon_{\mathrm{m}}''|\epsilon_{\mathrm{m}}|^{(m-3)/2}}{\epsilon_{\mathrm{d}}^{(m-1)/2}} \quad (d=2), \tag{5.55}$$

for $m > 1$.

For a Drude metal and $\omega \ll \omega_p$, from (5.54) and (5.55), we obtain

$$M_{n,m} \sim \epsilon_{\mathrm{d}}^{(1-n-m)/2}(a/\xi_{\mathrm{A}})^{2(n+m-1)}\left(\omega_p/\omega_\tau\right)^{n+m-1} \quad (d=2), \tag{5.56}$$

for $n + m > 1$ and $n > 0$, and

$$M_{0,m} \sim \epsilon_{\mathrm{d}}^{(1-m)/2}\left(\frac{\omega_p^{m-1}\omega_\tau}{\omega^{m}}\right) \quad (d=2), \tag{5.57}$$

for $m > 1$.

Note that for all moments the maximum in (5.56) and (5.57) is approximately the same (if $\xi_{\mathrm{A}} \sim a$ and $\rho \sim 1$), so that

$$M_{n,m}^{(\max)} \sim \epsilon_{\mathrm{d}}^{(1-n-m)/2}\left(\frac{\omega_p}{\omega_\tau}\right)^{n+m-1} \quad (d=2). \tag{5.58}$$

However, in the spectral range $\omega_p \gg \omega \gg \omega_\tau$, moments $M_{0,m}$ increase with the wavelength and the maximum is reached only at $\omega \sim \omega_\tau$, whereas the moments $M_{n,m}$ (with $n > 1$) reach this maximum at much shorter wavelengths (roughly, at $\omega \approx \tilde{\omega}_p/2$) and remain almost constant in the indicated spectral interval.

Figure 5.7 compares results of numerical and theoretical calculations for the field moments in two-dimensional silver semicontinuous films on glass. We see that there is excellent agreement between the scaling theory [formulas

(5.54) and (5.55)] and numerical simulations. To fit the data we used $\xi_A \approx 2a$. (Results of numerical simulations for $M_{0,4}$ are not shown since it was not possible to achieve reliable results in the simulations because of large fluctuations in values of this moment.)

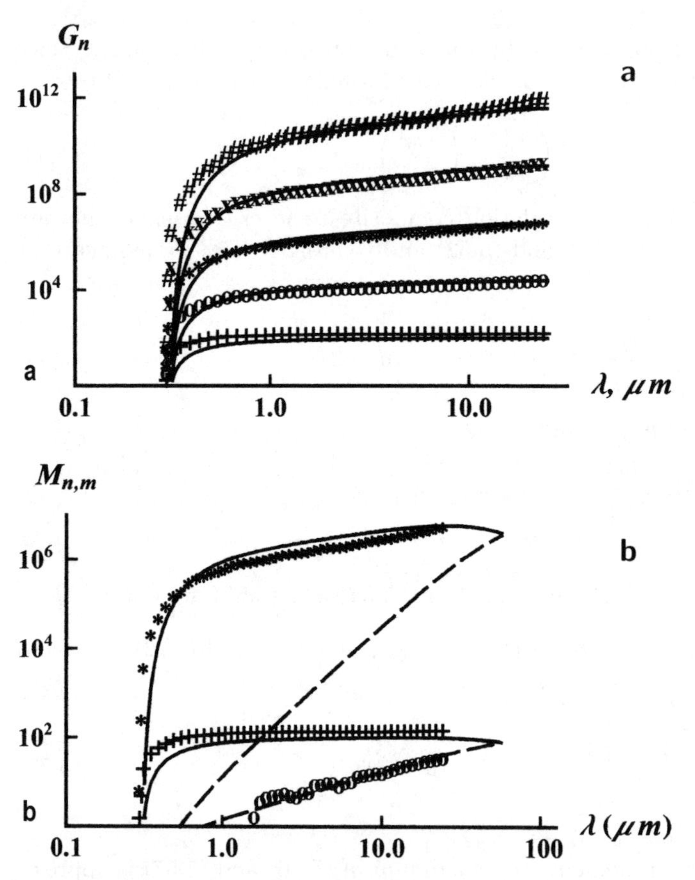

Fig. 5.7. (**a**) Average enhancement of the high-order field moments $G_n \equiv M_{n,0} = \langle |E/E^{(0)}|^n \rangle$ in a silver-glass two-dimensional random film as a function of the wavelength at $p = p_c$. Results of the numerical calculations for $n = 2, 3, 4, 5$ and 6 are represented by +, 0, *, x and # respectively. The solid lines describe G_n found from the scaling formulas. (**b**) Moments in a silver-glass random film as functions of wavelength at $p = p_c$: $M_{4,0} = \langle |E/E^{(0)}|^4 \rangle$ (*scaling formula* (5.56), *upper solid line; numerical simulations*, *); $M_{0,4} = \langle (E/E^{(0)})^4 \rangle$ (*scaling formula* (5.57), *upper dashed line*); $M_{2,0} = \langle |E/E^{(0)}|^2 \rangle$ (*scaling formula* (5.56), *lower solid line; numerical simulations* +); $M_{0,2} = \langle (E/E^{(0)})^2 \rangle$ (*scaling formula* (5.57), *lower dashed line; numerical simulations*, 0). The moment $M_{4,0}$, in particular, gives enhancement for Raman scattering and Kerr optical nonlinearity; moments $|M_{0,2}|^2$ and $|M_{0,4}|^2$ are enhancement factors for second and fourth harmonic generation, respectively

As discussed above, nonlinear optical processes are in general phase-dependent and proportional to a factor $|E|^n E^m$, i.e. they depend on the phase through the term E^m, and their enhancement is estimated as $M_{n,m} = \left\langle \left|E/E^{(0)}\right|^n \left(E/E^{(0)}\right)^m \right\rangle$. According to the above consideration, $M_{n,m} \sim M_{n+m,0}$, for $n \geq 1$. For example, enhancement of the Kerr-type nonlinearity $G_{\mathrm{K}} = M_{2,2} \sim G_{\mathrm{RS}} \simeq M_{4,0}$; it is frequency independent and estimated as $G_{\mathrm{K}} \sim \epsilon_{\mathrm{d}}^{-3/2} (a/\xi_{\mathrm{A}})^6 (\omega_p/\omega_\tau)^3$ (for a silver semicontinuous film on glass, $G_{\mathrm{K}} \sim 10^6$). For nearly degenerate four-wave mixing (FWM), the enhancement is given by $G_{\mathrm{FWM}} \sim |G_{\mathrm{K}}|^2 \sim |M_{2,2}|^2$ and can reach giant values up to $\sim 10^{12}$. The enhancement for mth harmonic generation, when $m\omega > \tilde{\omega}_p$ and $d = 2$, is estimated as $G_{m\mathrm{HG}} \sim |M_{0,m}|^2 \sim \epsilon_{\mathrm{d}}^{(1-m)} (\omega_p/\omega)^{2(m-1)} (\omega_\tau/\omega)^2 \gg 1$, i.e., in contrast to the processes with "photon subtraction", the moment depends on the frequency, in the range $\omega_p \gg \omega \gg \omega_\tau$, and reaches approximately the same level of enhancement as the processes with a photon subtraction, but only at $\omega \sim \omega_\tau$.

For three-dimensional metal-dielectric composites, with the dielectric constant of metal component estimated by the Drude formula, moments $M_{n,m} (n \neq 0)$ are also strongly enhanced for $\omega < \tilde{\omega}_p$, such that

$$
\begin{aligned}
M_{n,m}(\omega \ll \omega_p) &\sim (\xi_{\mathrm{A}}/a)^{3-2(n+m)} \left[\left(\frac{1}{\epsilon_{\mathrm{d}}} \frac{\omega_p^2}{\omega^2} \right)^{\nu/(t+s)} \left(\frac{\omega}{\omega_\tau} \right) \right]^{n+m-1} \\
&\sim (\xi_{\mathrm{A}}/a)^{3-2(n+m)} \left[\frac{\omega_p^{2/3} \omega^{1/3}}{\epsilon_{\mathrm{d}}^{1/3} \omega_\tau} \right]^{n+m-1} \qquad (d = 3), \qquad (5.59)
\end{aligned}
$$

where the relation $\nu/(t+s) \approx 1/3$ is used. For the three-dimensional case, the moments (and the field peaks E_{m}) achieve the maximum value at frequency $\omega_{\max} \approx 0.5\tilde{\omega}_p$ and then decrease with the wavelength; the maximum value is estimated as $M_{n,m} \sim (\xi_{\mathrm{A}}/a)^{3-2(n+m)} \left[(\epsilon_0/\epsilon_{\mathrm{d}})^{1/3} \tilde{\omega}_p/\omega_\tau \right]^{n+m-1}$.

It is interesting to note that the strong increase in enhancement with the order of optical nonlinearity can result in unusual situation when, for example, second-harmonic generation (SHG) is dominated by higher-order nonlinearity $\chi^{(4)}(-2\omega; \omega, \omega, -\omega, \omega)$, rather than being due to $\chi^{(2)}(-2\omega; \omega, \omega)$. This is because $M_{2,2}$, which is responsible for enhancement of the above $\chi^{(4)}(-2\omega; \omega, \omega, -\omega, \omega)$, can significantly exceed $M_{2,0}$ (e.g., by seven orders of magnitude at $\sim 1\,\mu$m). Another possible situation is when hyper-Raman scattering is as efficient as "conventional" Raman scattering. Also, we note that when higher-order nonlinearities compete with lower-order nonlinearities, bistable behavior can be obtained, which can be used in various applications in optoelectronics.

It is also important to note that local enhancements in the hot spots can be much larger than the average enhancements considered above. As follows from Figs. 5.3, 5.5 and 5.6, the local field intensities are enhanced up to

$|E/E^{(0)}|^2 \sim 10^5$; therefore, for the Kerr-type process and four-wave mixing, the local enhancements in the hot spots can be up to 10^{10} and 10^{20} respectively. With such a level of local enhancements, the nonlinear optical signals from single molecules and nanoparticles can be detected, which opens a fascinating possibility of local nonlinear spectroscopy with nanometer spatial resolution.

For the sake of simplicity, we assumed above that $p = p_c$. Now we estimate the concentration range $\Delta p = p - p_c$, where the above estimates for enhanced optical phenomena are valid. Although these estimates have been made for the percolation threshold $p = p_c$, they must also be valid in some vicinity of the threshold. The above speculations are based on finite-size scaling arguments, which hold provided that the scale l_r of the renormalized elements is smaller than the percolation correlation length $\xi \cong a(|p - p_c|/p_c)^{-\nu}$. At the percolation threshold, where the correlation length ξ diverges, our estimates are valid in a wide frequency range $\omega_\tau < \omega < \tilde{\omega}_p$ that includes the visible, near-infrared and mid-infrared spectral ranges for a typical metal. For any particular frequency within this interval, we estimate the concentration range Δp, where the giant field fluctuations occur, by equating the values of l_r from (5.48) and ξ. Thus we obtain the following relation:

$$|\Delta p| \le (\epsilon_d/|\epsilon_m|)^{1/(t+s)}. \tag{5.60}$$

For a two-dimensional semicontinuous metal film, the critical exponents are $s \approx t \approx \nu_p = 4/3$, and the above relation acquires the form

$$|\Delta p| \le (\epsilon_d/|\epsilon_m|)^{3/8}. \tag{5.61}$$

For a Drude metal, in the frequency range $\tilde{\omega}_p \gg \omega \gg \omega_\tau$, (5.61) can be rewritten as follows

$$\Delta p \le \epsilon_d{}^{3/8} (\omega/\omega_p)^{3/4}. \tag{5.62}$$

As follows from (5.62), the concentration range for the enhancement shrinks when the frequency decreases much below the plasma frequency. This result is in agreement with the computer simulations shown in Fig. 5.4.

We also note the fact that the problem considered here maps onto the Anderson transition problem brings about a new experimental means to study the long-standing Anderson problem by taking advantage of high intensities and coherence of the laser light and sub-wavelength resolution provided by the near-field optical microscopy recently developed. In particular, by studying the high-order field moments associated with various nonlinear optical processes and spatial distribution of the nonlinear signal, unique information can be obtained on the eigenfunction distribution for the Anderson problem.

5.3.4 Comparison of Surface-Enhanced Phenomena in Fractal and Percolation Composites

It is interesting to compare the enhancement of optical nonlinearities in fractals considered in Chap. 3 and in percolation systems. In both cases, localization of optical excitations in small areas with different local configurations leads to an inhomogeneous broadening of the spectra. In this case, enhancement for arbitrary nonlinearity can be estimated by the resonance value $|E_{\mathrm{res}}/E^{(0)}|^n$ multiplied by the fraction of the resonant modes f, i.e. by the fraction of particles involved in the resonance excitation.

In fractals, the fraction of resonance monomers is given by $\sim \delta \mathrm{Im}[\alpha(X)]$, where $\alpha_0 \equiv -(X + \mathrm{i}\delta)^{-1}$ is the polarizability of an individual particle (we set the radius of particles $R_{\mathrm{m}} = 1$) and $\alpha(X)$ is the average polarizability in a cluster. The quantity δ characterizes a homogeneous width of individual resonances (in the X space) forming the inhomogeneously broadened absorption spectrum $\mathrm{Im}[\alpha(X)]$ of fractals; δ^{-1} gives quality-factors of individual resonances. Spectral dependencies of X and δ are given in (3.30) and (3.31) respectively.

The resonance local field is estimated by $|E_{\mathrm{res}}/E^{(0)}| \sim 1/\delta$, as follows from (3.36), which is the exact solution. This result can be also obtain from the simple fact that the linear polarizability α_i of the ith monomer in a cluster experiences a shift of the resonance w_i because of interactions with other particles (where w_i is a real number, in the quasistatic approximation). Therefore the ith polarizability can always be represented as $\alpha_i = 1/(\alpha_0^{-1} + w_i) = [(w_i - X) - \mathrm{i}\delta]^{-1}$, where the shift w_i depends on interactions with all particles. In the limit of non-interacting particles, $w_i = 0$ and $\alpha_i = \alpha_0 = -(X+\mathrm{i}\delta)^{-1}$. For the resonance particles $w_i = X(\omega_r) \equiv X_r$ and $\alpha_r = \mathrm{i}/\delta$. The local field is related to the local polarizability as $E_i = \alpha_0^{-1}\alpha_i E^{(0)}$[see (3.33)], so that for $X \gg \delta$ we obtain for the resonance field $|E_{\mathrm{res}}| \sim (|X_r|/\delta)|E^{(0)}|$. In the optical spectral range $|X_{\mathrm{r}}| \sim 1$, and $|E_{\mathrm{res}}| \sim \delta^{-1}|E^{(0)}|$.

The quantity $\mathrm{Im}[\alpha(X)]$ represents the ensemble average absorption (extinction) and can be expressed in terms of the imaginary part of the effective dielectric function ϵ_{e}'' by $\epsilon_{\mathrm{e}}'' = 4\pi p \epsilon_{\mathrm{d}}(4\pi R_{\mathrm{m}}^3/3)^{-1}\mathrm{Im}[\alpha(X)]$, where p is the metal filling factor [cf. (1.4)]; again, setting $R_{\mathrm{m}} = 1$ we obtain $\mathrm{Im}[\alpha(X)] = (3p\epsilon_{\mathrm{d}})^{-1}\epsilon_{\mathrm{e}}''$. The above relation between ϵ_{e} and $\mathrm{Im}\alpha(X)$ assumes that $p \ll 1$, as in fractals; still, to perform comparative estimates we set below $p = p_{\mathrm{c}} \sim 1/d$.

Now we consider a two-dimensional system near percolation, where the exact formula $\epsilon_{\mathrm{e}} = \sqrt{\epsilon_{\mathrm{m}}\epsilon_{\mathrm{d}}}$ is available (recall also that $\nu \approx t \approx s$, in this case). Then, for the fraction of metal particles involved in the resonance excitation, we obtain [see also (3.30) and (3.31)] at $\tilde{\omega}_p \gg \omega \gg \omega_\tau$

$$f^{\mathrm{frac}} \sim \delta \mathrm{Im}[\alpha(X)] \sim \frac{\epsilon_{\mathrm{m}}''}{|\epsilon_{\mathrm{m}}|^2}|\epsilon_{\mathrm{m}}|^{1/2}\epsilon_{\mathrm{d}}^{1/2} \sim \frac{\epsilon_{\mathrm{m}}''\epsilon_{\mathrm{d}}^{1/2}}{|\epsilon_{\mathrm{m}}|^{3/2}}. \tag{5.63}$$

For a percolation system, the same resonance fraction for the two-dimensional case is estimated as (see Sect. 5.3.3)

$$f^{\mathrm{per}} \sim \frac{n(l_{\mathrm{r}})}{(\xi_{\mathrm{e}}/a)^2} \sim \frac{(l_{\mathrm{r}}/a)}{(\xi_{\mathrm{e}}/a)^2} \sim \left(\frac{|\epsilon_{\mathrm{m}}|}{\epsilon_{\mathrm{d}}}\right)^{1/2} \frac{\epsilon_{\mathrm{m}}''}{|\epsilon_{\mathrm{m}}|}\left(\frac{\epsilon_{\mathrm{d}}}{|\epsilon_{\mathrm{m}}|}\right) \sim \frac{\epsilon_{\mathrm{m}}''\epsilon_{\mathrm{d}}^{1/2}}{|\epsilon_{\mathrm{m}}|^{3/2}}, \tag{5.64}$$

which coincides with the result obtained above for fractals. (Hereafter, we assume, for simplicity, that $\xi_A \sim a$.) Thus, the fraction of resonance monomers predicted by fractal theory (in the limit corresponding to a percolation system) is the same as the one following from percolation theory.

We compare now the resonance fields and the high-order field moments predicted by the two theories. Fractal theory, in the limit of a percolation system, predicts [see also (3.47)–(3.49)] that

$$\frac{E_{\rm res}^{\rm frac}}{E^{(0)}} \sim \delta^{-1} \sim \frac{|\epsilon_{\rm m}|^2}{\epsilon_{\rm m}''\epsilon_{\rm d}}, \tag{5.65}$$

for the maximum fields, so that the moment $M_{k,m} \equiv G_n$ (where $n = k + m$ and $k > 1$) is given by

$$G_n^{\rm frac} \sim \frac{\delta {\rm Im}[\alpha(X)]}{\delta^n} \sim \frac{|\epsilon_{\rm m}|^{2n-3/2}}{\epsilon_{\rm d}^{n-1/2}(\epsilon_{\rm m}'')^{n-1}}. \tag{5.66}$$

The first relation in formula (5.66) reproduces, in particular, the exact result for the second moment given by (3.46).

Percolation theory for the two-dimensional case gives [see (5.49)]

$$\frac{E_{\rm res}^{\rm per}}{E^{(0)}} \equiv \frac{E_{\rm m}}{E^{(0)}} \sim \frac{(l_{\rm r}/a)}{\kappa} \sim \left(\frac{|\epsilon_{\rm m}|}{\epsilon_{\rm d}}\right)^{1/2}\left(\frac{|\epsilon_{\rm m}|}{\epsilon_{\rm m}''}\right) \sim \frac{|\epsilon_{\rm m}|^{3/2}}{\epsilon_{\rm d}^{1/2}\epsilon_{\rm m}''}. \tag{5.67}$$

For moments $M_{k,m} = G_n$, with $n = k + m$, the percolation scaling theory predicts [see (5.52)]

$$G_n^{\rm per} \sim \left(\frac{E_{\rm m}}{E^{(0)}}\right)^n \frac{n(l_{\rm r})}{(\xi_{\rm e}/a)^2} \sim \left(\frac{l_{\rm r}/a}{\kappa}\right)^{n-1} \sim \left(\frac{|\epsilon_{\rm m}|}{\epsilon_{\rm d}}\right)^{(n-1)/2}\left(\frac{|\epsilon_{\rm m}|}{\epsilon_{\rm m}''}\right)^{n-1}$$

$$\sim \epsilon_{\rm d}^{(1-n)/2}\left(\frac{|\epsilon|^{3/2}}{\epsilon_{\rm m}''}\right)^{n-1}. \tag{5.68}$$

By comparing (5.65) with (5.67), and (5.66) with (5.68), we obtain the following relations:

$$\frac{E_{\rm res}^{\rm frac}}{E_{\rm res}^{\rm per}} \sim \left(\frac{|\epsilon_{\rm m}|}{\epsilon_{\rm d}}\right)^{1/2} \sim l_{\rm r}/a \tag{5.69}$$

for the field peaks, and

$$\frac{G_n^{\rm frac}}{G_n^{\rm per}} \sim \left(\frac{|\epsilon_{\rm m}|}{\epsilon_{\rm d}}\right)^{n/2} \sim (l_{\rm r}/a)^n \tag{5.70}$$

for the high-order field moments.

Thus, we can say that, while the fraction of the resonance monomers excited by the applied field in both fractal and percolation composites is the same, the magnitude of the resonance fields in fractals is larger by the factor $l_{\rm r}/a$ leading to nth order moments larger by the factor $(l_{\rm r}/a)^n$. This can be explained as follows. Fractals consist mostly of dielectric voids, since the volume fraction filled by metal is very small and tends to zero with increasing size of the fractal. The maximum fields are localized at the metal-vacuum boundaries at the tips of various metal branches in a fractal. In a percolation system, in contrast, each field maximum is actually stretched out into a chain of smaller peaks along a dielectric gap of length $l_{\rm r}$ between neighboring metal clusters that resonate at the applied field frequency. By comparing the calculated second moments (see Figs. 3.6 and 5.4) and fourth moments (see Figs. 3.14b and 5.7) in silver fractals and semicontinuous films, we see that the high-order field moments in fractals significantly exceed the corresponding moments of percolation composites, with the ratio between the two roughly following the above estimate.

Similar estimates for three-dimensional systems, where the effective dielectric function of a random composite at the percolation is given by $\epsilon_{\rm m}(\epsilon_{\rm d}/\epsilon_{\rm m})^{t/(t+s)}$, show that the fraction of the resonance monomer can still be approximated by the formula (5.63), whereas the resonance fields in fractals exceeds those in a three-dimensional percolation system by a larger factor, $l_{\rm r}^2$, than in the two-dimensional case. In the general case, the ratio of the moments can be written as

$$\frac{G_{\rm n}^{frac}}{G_n^{\rm per}} \sim \left[(l_{\rm r}/a)^{d-1}\right]^n , \tag{5.71}$$

where $l_{\rm r}/a \sim |\epsilon_{\rm m}/\epsilon_{\rm d}|^{1/2}$ for $d = 2$, and $l_{\rm r}/a \sim |\epsilon_{\rm m}/\epsilon_{\rm d}|^{1/3}$ for $d = 3$ (we used the approximation $\nu/(t+s) \approx 1/2$, for $d = 2$, and $\nu/(t+s) \approx 1/3$, for $d = 3$). We can say that the larger enhancement in fractals is related to their fluctuation nature, i.e. to the fact that fractals have a mean density that tends to zero. In contrast, percolation systems are on average homogeneous, and the field fluctuations, being very large, are still less than in fractals, especially in the three-dimensional case.

We should note that the above comparison cannot, strictly speaking, be used when $\langle E\rangle \sim (X/\delta)\times\delta{\rm Im}[\alpha(X)] \sim X{\rm Im}[\alpha(X)]$ is much larger than unity. In a percolation system, however, the effective dielectric function $\epsilon_{\rm e}$ can be large in the infrared part of the spectrum such as to violate the approximation. Still, the arguments used in the fractal theory allow the obtaining of some useful estimates for a percolation composite, at least for the visible part of the spectrum.

Also, it is important to note that enhancement for fractals is defined above per one metal particle. The enhancement per unit volume includes the additional factor $p \sim (R_{\rm c}/R_{\rm m})^{D-d}$, which is small in fractals of large $R_{\rm c}$. (For a percolation system, the difference in definitions does not matter if $p \sim 1$.) Thus the average enhancement per unit volume in a percolation composite can exceed that for fractals, if the latter have very large sizes. In this sense, it might be better to use a composite medium, formed by many fractals of a reasonable size (say $\sim 1\,\mu$m) that are distributed in a host medium so that the average distance between the fractals is of the order of their size.

5.4 Some Experimental Results

5.4.1 Near-Field Nano-Optics

In this section we consider some experimental studies related to the theory that has been developed above.

In [168] the imaging and spectroscopy of localized optical excitations in gold-on-glass percolation films was performed, using scanning near-field optical microscopy (SNOM). The SNOM probe used was an apertureless tip made of a tungsten wire etched by electrochemical erosion. The radius of curvature of the tip, measured by scanning electron microscopy, was about 10 nm, which provided the very high spatial resolution needed to image the localized optical modes. The tip was set above a sample attached to an (x, y) horizontal piezoelectric stage. The tip was the bent end of the tungsten wire, used as a cantilever connected to a twin-piezoelectric transducer that could excite it perpendicularly to the sample surface. The frequency of vibrations ($\sim$ 5kHz) was close to the resonant frequency of the cantilever and its amplitude was about 100 nm. In the tapping mode, the tip vibrated above the sample as in atomic force microscope (AFM). Detection of the vibration amplitude was made by a transverse laser-diode probe beam focused at the lever arm. A feedback system, including a piezoelectric translator attached to the twin-piezoelectric transducer (needed for the tip vibrations), kept the vibration amplitude constant during the sample scanning. The detection of the feedback voltage applied to the piezoelectric translator gave a topographical image of the surface, i.e. the AFM signal that can be taken simultaneously with the SNOM signal. A first microscope objective focused the light of a tunable cw Ti:Sapphire laser on the bottom surface of the sample. The signal collection was axially symmetric above the sample and made by a second microscope objective. The transmitted light passing through the microscope was then sent to a photomultiplier. Samples of semicontinuous metal films were prepared by depositing gold thin films on a glass substrate at room temperature under ultrahigh vacuum (10^{-9} Torr). The tip vibration modulated the near-zone field on a sample surface and lock-in detection of the collected light at the tip's vibrational frequency allowed the detection of the locally modulated field.

For the visible and near-IR parts of the spectrum, the tungsten tip ($n \approx 3.5 + 2.8i$) was found not to have any resonances, so that its polarizability is much less than the polarizability of the resonance-enhanced plasmon oscillations of the film. Because of this, perturbations in the field distribution introduced by the tip are relatively small.

Figures 5.8a and 5.8b show experimental and calculated near-field images at the surface of a percolation gold-glass film, for different wavelengths (experimental and simulated samples are, on average, similar but they differ locally). Note that optical excitations are localized in both horizontal and vertical directions. Our estimations show that the signal becomes negligible at the tip-sample separation of ~ 10 nm. This means that in the tapping mode, with oscillation amplitude 100 nm, the average detected signal is strongly decreased, by a factor of roughly 1/10 to 1/100. To approximate this and the finite size of the tip in simulations of [168], the tip-sample separation was set at 10 nm and the collected signal was averaged over the area $50 \times 50\,\mathrm{nm}^2$.

The near-field images observed are in qualitative agreement with the theoretical predictions and numerical simulations. The optical excitations of a percolation film are localized in ~ 100 nm-sized areas, significantly smaller than λ. The peak amplitudes and their spatial separations increase with λ, in accordance with theory. For SNOM detection, the experimentally observed and simulated enhancement of the local field intensity is 10 to 10^2. However, according to Fig. 5.3, the local field intensity right on the surface of the film (which can be probed, for example, by surface-adsorbed molecules) is larger than the averaged signal detected by the vibrating SNOM tip by roughly two orders of magnitude. Note that both observed and calculated near-field images are λ-dependent: even a small change of wavelength $\Delta\lambda \sim \lambda/50$ results in dramatically different field distributions, as shown in Fig. 5.8.

In [168] the near-field spectroscopy of percolation films was also performed, by parking an SNOM tip at different points of the surface and varying the wavelength. This local nano-spectroscopy allows the determination of the local resonances of nm-sized areas right underneath the tip; the nanostructures at different points resonate at different λ, leading to different local near-field spectra. The spectra characterize λ-dependence of the field hot spots associated with the localized sp modes.

The experimental and calculated spectra [168] shown in Figs. 5.8c and 5.8d are qualitatively similar to those detected for fractals (see Fig. 3.10). The spectra consist of several peaks $\sim$10 nm in width, and they depend markedly on spatial location of the point where the near-field tip is parked. As shown in [168], even as small a shift in space as 100 nm results in significantly different spectra, which is strong evidence of the surface plasmon localization. We note that, for continuous metal (or dielectric) films, neither sub-λ hot spots nor their local spectra can be observed, because in this case optical excitations are delocalized.

In [169], local nano-photomodification of percolation films was also observed. After the irradiation of a percolation film by a pump above a certain threshold, the local nm-scale configuration in the resonant hot spot experiences restructuring through particle sintering. As a result, after photomodification, the configuration might be no longer in resonance with the imaging field, and the local field-intensity significantly decreases in the area of photomodification. A similar effect was obtained for fractal particle aggregates (see Fig. 3.19).

Thus, the near-field imaging and spectroscopy of random metal-dielectric films near percolation suggests the localization of optical excitations in small

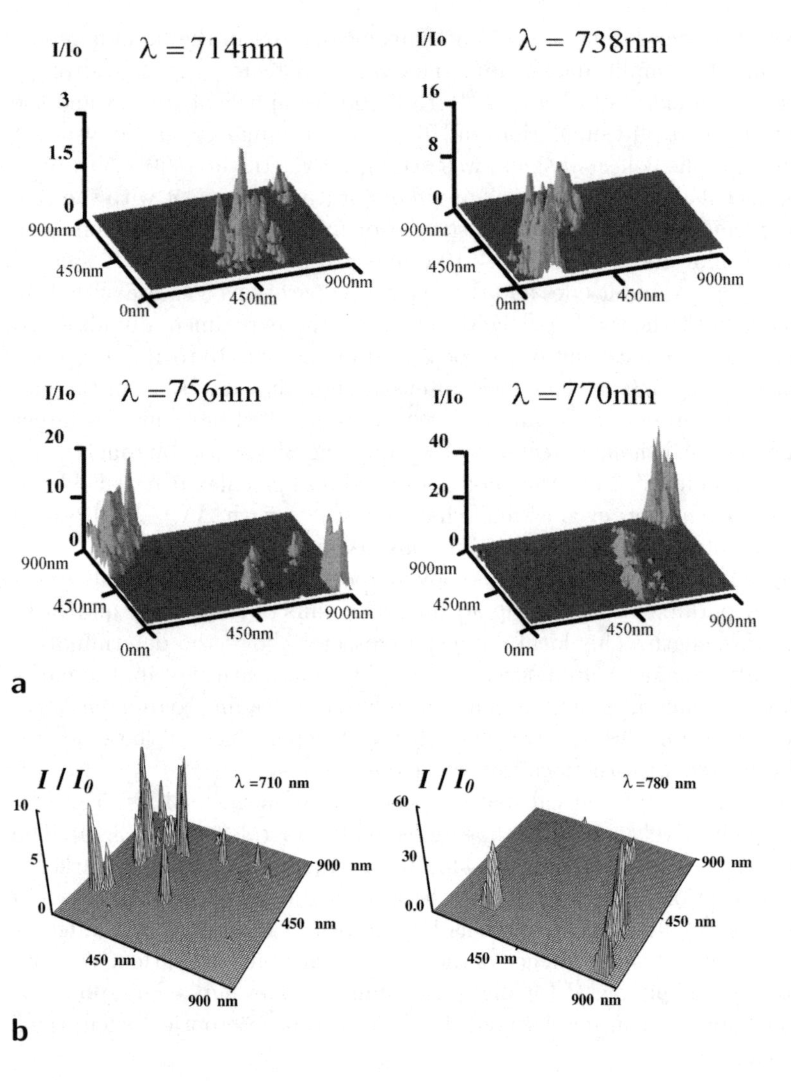
I/Io
λ = 714nm
3
1.5
0
900nm
450nm
0nm
450nm
900nm
I/Io
λ = 738nm
16
8
0
900nm
450nm
0nm
450nm
900nm
I/Io
λ = 756nm
20
10
0
900nm
450nm
0nm
450nm
900nm
I/Io
λ = 770nm
40
20
0
900nm
450nm
0nm
450nm
900nm
a
I / I0
λ = 710 nm
10
5
0
900 nm
450 nm
450 nm
900 nm
I / I0
λ = 780 nm
60
30
0.0
900 nm
450 nm
450 nm
900 nm
b

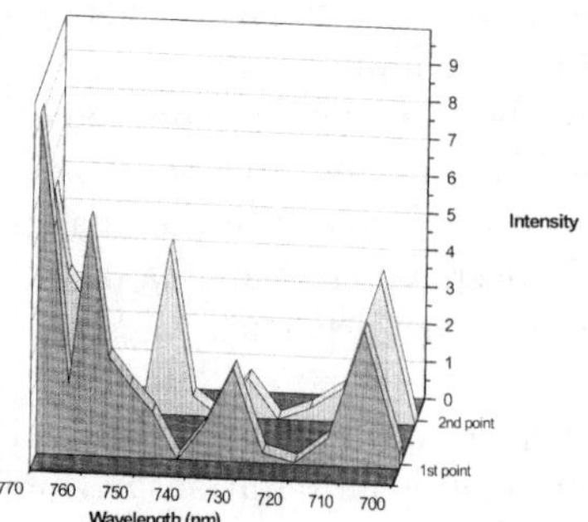
Intensity
2nd point
1st point
Wavelength (nm)
770 760 750 740 730 720 710 700
c

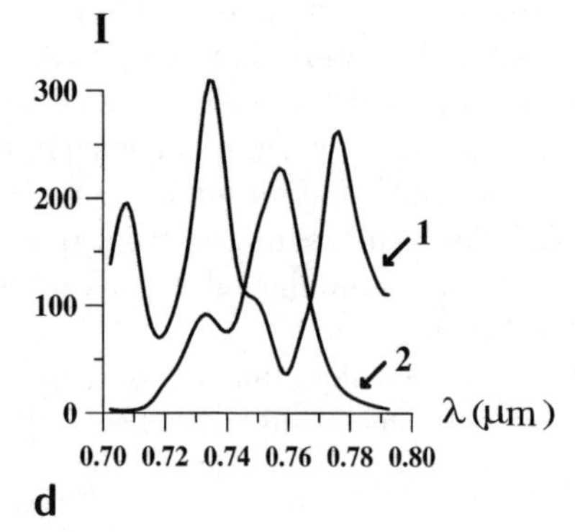
I
300
200
100
0
1
2
λ (μm)
0.70 0.72 0.74 0.76 0.78 0.80
d

nm-scale hot spots, as predicted by theory. The observed pattern of the localized modes and their spectral dependences are in agreement with theoretical results and numerical simulations described in previous sections of this book. The hot spots of a percolation film represent very large local fields (fluctuations); spatial positions of the spots strongly depend on the light frequency. Near-field spectra observed and calculated at various points of the surface consist of several spectral resonances, whose spectral locations depend on the probed site of the sample. All these features are only observable in the near zone. In the far zone, images and spectra are observed in which the hot spots and the spectral resonances are averaged out. The local field enhancement is large, which is especially important for nonlinear processes of the nth order, which are proportional to the enhanced local fields to the nth power. This opens up a fascinating possibility for *nonlinear* near-field spectroscopy of single nanoparticles and molecules.

5.4.2 Surface-Enhanced Raman Scattering

As shown above, the scaling approach for surface-enhanced optical phenomena is valid approximately in some interval $\Delta p = p - p_{\mathrm{c}}$ in the vicinity of p_{c}, when the size l_{r} is smaller than or equal to the percolation correlation length, $\xi \cong a|\Delta p|^{-\nu}$. This, in particular, allows an estimation for the concentration range where SERS occurs: $\Delta p \leq \Delta^* = (\epsilon_{\mathrm{d}}/|\epsilon_{\mathrm{m}}|)^{1/(t+s)}$. Based on this estimate, the following scaling formula for the enhancement of Raman scattering was proposed [149]:

$$G_{\mathrm{RS}}(p,\omega) = G^0_{\mathrm{RS}}(p_c,\omega)F(\Delta p/\Delta^*). \tag{5.72}$$

To find the scaling function $F(x)$, the quantity G_{RS} was calculated for various ω and p. For very different frequencies used in these simulations, the results collapse onto one curve $F(x)$ shown in Fig. 5.9a. This function has its first (small) maximum below p_{c}; at $p = p_{\mathrm{c}}$, the function is $F(0) \approx 1$; then $F(x)$ has another larger maximum at $p > p_{\mathrm{c}}$, and it finally vanishes, at $|\Delta p| > \Delta^*$.

This double-peak behavior of $F(x)$ is related to the evolution with p of the plasmon resonances discussed above (see also Fig. 5.2). To explain the minimum near $p = p_{\mathrm{c}}$, we note the following. The localization radius ξ_{A} of the eigenstates Ψ_n, with most important eigenvalues $\Lambda_n \approx 0$, decreases

---

Fig. 5.8. Experimental (**a**) and calculated (**b**) SNOM images of the localized optical excitations in a percolation gold-on-glass film for different wavelengths λ. Note that the actual field intensities right on the film surface [that can be probed, for example, by surface-adsorbed molecules; see Fig. 5.3)] are by two to three orders of magnitude larger than those detected by a finite-size tip of SNOM. Experimental (**c**) and calculated (**d**) near-field spectra at different spatial locations (100 nm apart) of the film. Arbitrary intensity units are used

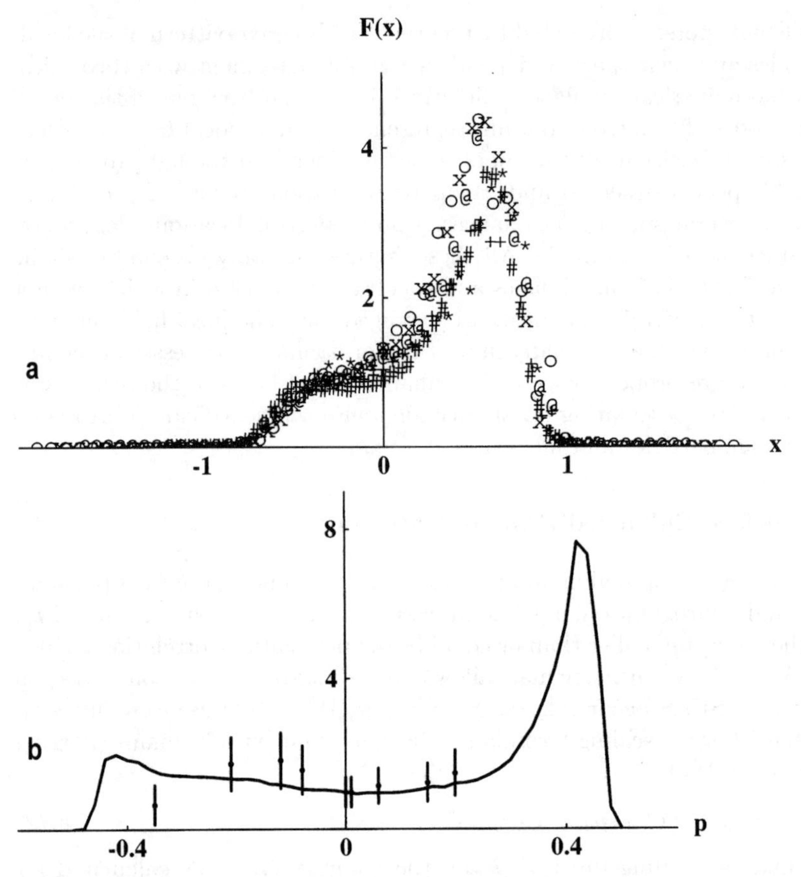

Fig. 5.9. (**a**) SERS scaling function $F(x)$ vs $x = (p - p_c)/\Delta^*$, for different wavelengths: $\lambda = 0.9\,\mu m$ (#), $\lambda = 1.1\,\mu m$ (+), $\lambda = 1.3\,\mu m$ (∗), $\lambda = 1.5\,\mu m$ (@), $\lambda = 1.7\,\mu m$ (×), and $\lambda = 1.9\,\mu m$ (*o*). (**b**) The normalized SERS $G_{RS}(p)/G_{RS}(p = p_c)$ as a function of the metal concentration, $\Delta p = p - p_c$, on a silver semicontinuous film (*solid line, theoretical calculations; points, experimental data*)

when we shift from $p = p_c = 1/2$ toward $p = 0$ or $p = 1$. This is because the eigenvalue $\Lambda = 0$, shifts (when moving away from $p = p_c$) from the center of the eigenvalue distribution to its tails, where localization is stronger [157]. According to (5.47) and (5.54), the enhancement factor $G_{RS} \approx M_{4,0}$ increases when moving either side from the point $p = p_c$, since ξ_A becomes smaller there.

In agreement with theoretical consideration, SERS double-maximum behavior in the p dependence was observed in experiments performed by Gadenne et al. [149]. The results of numerical simulations and experimental data are shown in Fig. 5.9b. In these experiments, SERS was detected for

Raman-active molecules of Zinc TetraPhenyPropilene (ZnTPP) adsorbed on a silver semicontinuous film.

5.4.3 Kerr Optical Nonlinearity and White-Light Generation

As shown above, enhancement of the Kerr optical nonlinearity is given by $G_{\mathrm{K}} = M_{2,2}$. Fig. 5.10 shows the results of numerical simulations for G_{K} as a function of the metal filling factor p, for $d = 2$. The plot, as above for G_{RS}, has a two-peak structure; however, in contrast to G_{RS}, the dip at $p = p_{\mathrm{c}}$ is much stronger and is proportional to κ. This implies that, at $p = p_{\mathrm{c}}$, the enhancement is actually given by $G_{\mathrm{K}} \sim \kappa M_{2,2}$, where $M_{2,2}$ was found above.

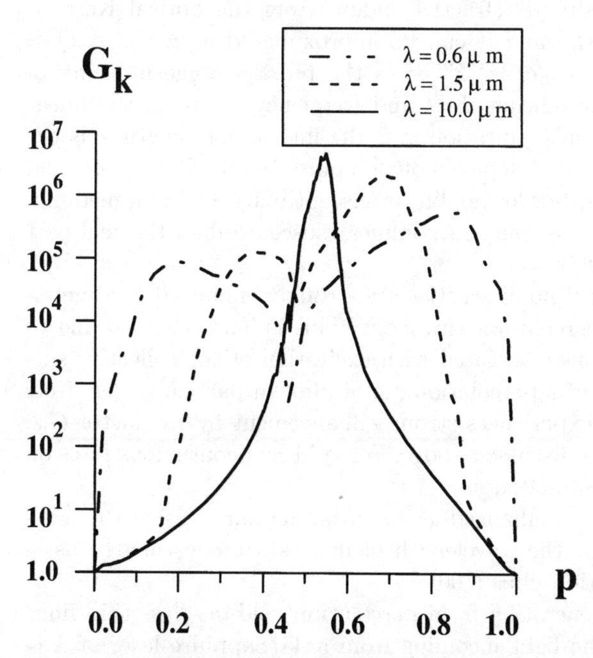

Fig. 5.10. Average enhancement of the Kerr optical nonlinearity $G_{\mathrm{K}} = M_{2,2}$ on a silver semicontinuous film as a function of the metal concentration p for three different wavelengths

This result might be a consequence of the special symmetry of a self-dual system at $p = p_{\mathrm{c}}$. Formally, it could happen if the leading term in the power expansion of $M_{2,2}$ over $1/\kappa$ cancels out because of the symmetry [see the discussion following (5.46)] and the next term must be taken into account. When this symmetry is somehow broken, e.g. by slightly moving away from the point $p = p_{\mathrm{c}}$, the enhancement G_{K} increases and becomes $G_{\mathrm{K}} \sim M_{2,2} \sim G_{\mathrm{RS}}$, as seen in Fig. 5.10.

The fact that the minimum at $p = p_c$ is much less for SERS than for the Kerr process can be related to the fact that the latter is a phase-sensitive effect.

The Kerr optical nonlinearity results particularly in nonlinear correction of the refractive index. It can result, for example, in self-modulation and spectral continuum generation, considered below.

Spectral continuum (also referred to as supercontinuum) generation, discovered in 1970, has now been demonstrated in a wide variety of solids, liquids, and gases [170]. Self-phase modulation, four-wave mixing, and plasma production (for the short-wavelength part of the spectrum) are most commonly invoked to explain continuum generation. According to a theory of self-phase modulation, the frequency-relative shift is proportional to the rate of change of the laser-induced refractive index (from the optical Kerr effect), such that $\Delta n = n_2 I_0$, and it can be approximated as $\Delta\omega/\omega \equiv Q \approx \beta z n_2 I_0/(c\tau)$, where $I_0 = (n_0 c/2\pi)|E^{(0)}|^2$ is the pulse average intensity of the applied field of the amplitude $E^{(0)}$ and frequency ω, n_0 is the linear refractive index, τ is the pulse duration, z is the interaction length, c is the speed of light, and prefactor β depends on the pulse temporal envelope and is ~ 1. In terms of the third-order nonlinear susceptibility $\chi^{(3)}$ of a medium, $n_2 \approx (12\pi^2/n_0^2 c)\chi^{(3)}$ (for the complex nonlinear susceptibility, the real part of $\chi^{(3)}$ is responsible for n_2).

As shown above, optical nonlinearities are strongly enhanced in a metal-dielectric film near the percolation threshold. This enhancement is due to giant local-field fluctuations associated with excitation of the collective surface plasmon (sp) modes of a percolation metal film. In particular, the Kerr optical nonlinearity $\chi^{(3)}$ experiences strong enhancement by the factor G_K, which can be very large as discussed above. For gold semicontinuous films on a glass substrate, G_K is estimated as $\sim 10^5$.

The value of the surface-enhanced $\chi^{(3)}$ can be measured using the four-wave mixing technique; at the wavelength of interest, it is estimated as $\sim 10^{-7}$ esu for gold percolation films [100].

Spectral continuum generation from percolation gold-on-glass thin films was observed in [171]. The light incoming from a Ti/Sapphire laser at $\lambda \sim 0.79\,\mu m$ and repetition rate of 76 MHz was focused on an area $\sim 1\,\mu m^2$ on the surface of the film. The pulse had a duration $\tau \sim 200$ fs. The white light generation was observed for energies as low as $\sim 10^{-11}$J. Samples of percolation gold films were prepared by depositing gold thin films on a glass substrate at room temperature under ultrahigh vacuum (10^{-9} Torr).

Figure 5.11 shows spectral dependence of the relative intensity for the generated continuum at the laser radiation with $\lambda = 0.79\,\mu m$ and intensity $I \approx 10^{11}\,W/cm^2$. In experiments [171], the continuum was detected in a large spectral range, from approximately 400 nm to 650 nm, which corresponded to the apparatus cutoffs, so that the actual generated range was probably much broader. The observed spectral distribution is relatively flat because the sp

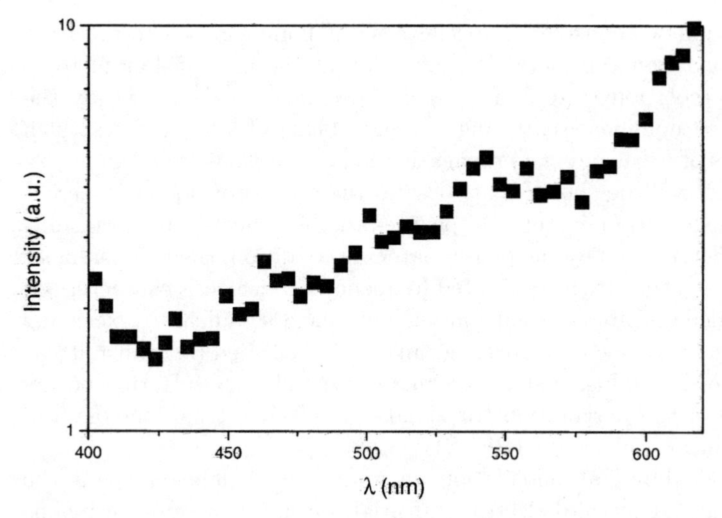

Fig. 5.11. Relative intensities of spectral continuum generated from random gold-on-glass film ($p = p_c$) at a wavelength of incident light $\lambda = 0.79\,\mu m$ and a laser pulse intensity and duration $I = 10^{11} W/cm^2$ and $\tau \approx 200\,fs$ respectively

eigenmodes of a percolation metal film cover a large spectral interval that includes the visible and near-to-mid infrared parts of the spectrum, leading to the surface-enhanced $\chi^{(3)}$ in this broad spectral interval.

The value $\chi^{(3)} \sim 10^{-7}$ esu for the nonlinear susceptibility of percolation gold films corresponds to the nonlinear refraction factor $n_2 \sim 10^{-9}\,cm^2/W$. Using this value of n_2, we find that the factor Q becomes of the order of unity, as required for supercontinuum generation at laser intensity $I \sim 10^{11}\,W/cm^2$, which is two to three orders of magnitude less that the intensity typically used to obtain white-light generation in media with no surface enhancement [170], and is also in agreement with experimental observations [171].

We can attribute this three-order-of-magnitude decrease of the intensity needed for supercontinuum generation to the surface enhancement of the Kerr nonlinearity resulting from excitation of the collective sp modes of a percolation film.

5.5 Percolation-Enhanced Nonlinear Scattering

As shown above, the local fields are strongly enhanced in the visible and infrared spectral ranges for a metal-dielectric film. Nonlinear optical processes of the nth order are proportional to $E^n(\mathbf{r})$, and therefore the enhancement can be especially large, for the resonance groups of particles, where the local field experiences giant fluctuations. The strong spatial fluctuations of the "nonlinear" field source, $\propto E^n(\mathbf{r})$, can result in giant nonlinear scattering from a composite material.

In this section we consider a special case of nonlinear scattering from a two-dimensional random metal-dielectric film at the metal filling-factor p close to the percolation threshold, $p \approx p_c$; specifically, we shall study the surface-enhanced nonlinear scattering of a light beam of frequency ω, which results in waves of frequency $n\omega$ propagating in all directions from the metal-dielectric film. It will be shown that, besides the above considered coherent signal of the nth harmonic that propagates in the reflected (or transmitted) direction determined by the phase-matching condition, there is nonlinear diffusive-like scattering at the generated frequency $n\omega$, which is characterized by a broad-angled distribution and giant enhancement significantly exceeding that for the nth harmonic beam propagating in the reflected (or transmitted) direction. Since this diffusive strongly-enhanced nonlinear scattering occurs in the vicinity of the percolation threshold, we refer to it as "percolation-enhanced nonlinear scattering" (PENS).

As described above, at percolation an infinite metal cluster spans the entire sample and a metal-dielectric transition occurs in a semicontinuous metal film. Optical excitations of the self-similar fractal clusters formed by metal particles near p_c result in giant scale-invariant field fluctuations. This makes the PENS considered here special in a family of similar phenomena of harmonic generation from "conventional" smooth [172] and rough [173] metal surfaces.

We assume that a semicontinuous metal film is covered by some layer possessing nonlinear conductivity $\sigma^{(n)}$ that results in nHG (e.g., it can be a layer of nonlinear organic molecules, semiconductor quantum dots, or a quantum well on top of the percolation film). The local electric field $\mathbf{E}_\omega(\mathbf{r})$ induced in the film by the external field $\mathbf{E}_\omega^{(0)}$ generates in the layer an $n\omega$ current $\sigma^{(n)}\mathbf{E}_\omega E_\omega^{n-1}$ [174]. This nonlinear current in turn interacts with the film and generates the "seed" $n\omega$ electric field, with amplitude $\mathbf{E}^{(n)} = \sigma^{(n)} E_\omega^{n-1}\mathbf{E}_\omega/\sigma^{(1)}$, where $\sigma^{(1)}$ is the *linear* conductivity of the nonlinear layer at frequency $n\omega$. The electric field $\mathbf{E}^{(n)}$ can be thought of as an inhomogeneous external field exciting the film at $n\omega$ frequency. The nHG current $\mathbf{j}^{(n)}$ induced in the film by the "seed" field $\mathbf{E}^{(n)}$ can be found in terms of the nonlocal conductivity tensor $\widehat{\Sigma}(\mathbf{r},\mathbf{r}')$ that relates the applied (external) field at point $\mathbf{r}'$ to the current at point $\mathbf{r}$, $j_\beta^{(n)}(\mathbf{r}) = \int \Sigma_{\beta\alpha}^{(n)}(\mathbf{r},\mathbf{r}')E_\alpha^{(n)}(\mathbf{r}')\,\mathrm{d}\mathbf{r}'$, where $\Sigma_{\beta\alpha}^{(n)}$ is the conductivity tensor at frequency $n\omega$ and the integration is over the entire film area (see Sects. 5.2.1 and 5.2.2). The Greek indices take values $\{x, y\}$ and summation over repeated indices is implied. It is the current $\mathbf{j}^{(n)}$that eventually generates the nonlinear scattered field at frequency $n\omega$.

Using the numerical technique described above, we can calculate the spatial distribution of the nonlinear fields, similar to those shown in Figs. 5.3, 5.5 and 5.6. As an example, Fig. 5.12 shows the normalized real part of the 3ω local field $\mathrm{Re}\left[E^2(\mathbf{r})E_x(\mathbf{r})\right]/\left|E^{(0)}\right|^3$ in a two-dimensional silver-on-glass film at $p = p_c$ and $\lambda = 1.5\,\mu\mathrm{m}$ (for the ϵ_m of silver, the Drude dielectric function, with plasma frequency $\omega_p = 9.1$ eV and relaxation rate $\omega_\tau = 0.021$ eV, were

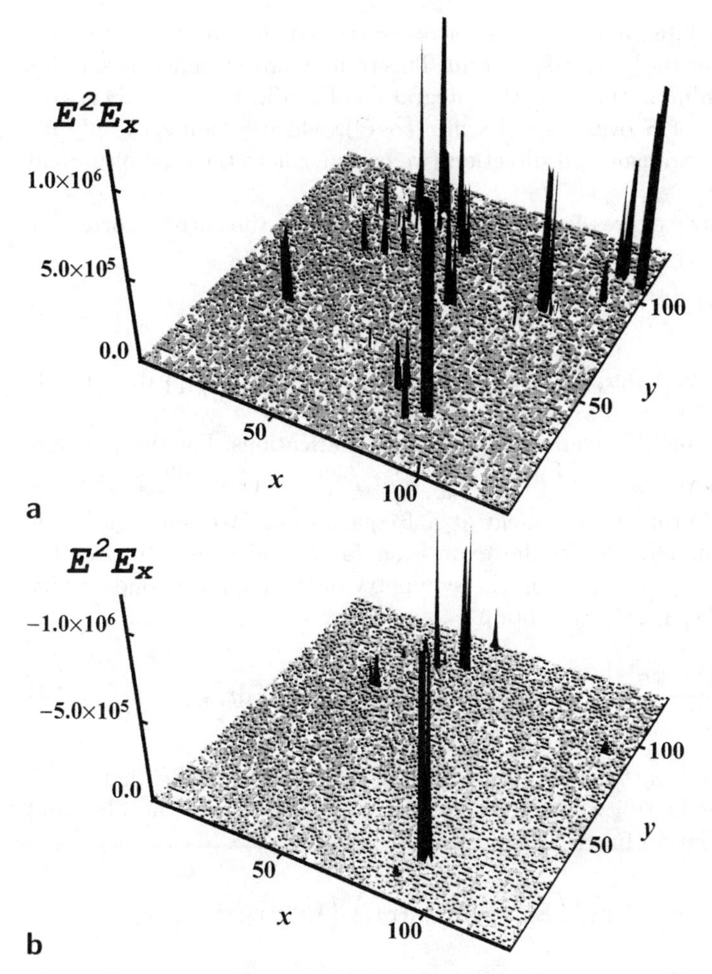

Fig. 5.12. Distribution of the x component of the nonlinear local field $\mathrm{Re}\left[E^2(\mathbf{r})E_x(\mathbf{r})\right]$, with positive (**a**) and negative (**b**) values. The applied field $E^{(0)} = 1$

used). As seen in Fig. 5.12, the fluctuating 3ω fields form a set of sharp peaks, looking up and down and having magnitudes $\sim 10^6$. Such huge fluctuations of the local fields are anticipated to trigger PENS at 3ω (and other harmonics), which is characterized by a broad-angled distribution.

By using the standard approach of scattering theory [155] and assuming that the incident light is unpolarized, we obtain that the integral scattering in all directions, except for the specular one (which is responsible for the coherent nth harmonic generation considered in the previous sections), is given by $S = \left(4k^2/3c\right) \int (\left\langle j_\alpha^{(n)}(\mathbf{r}_1)\, j_\alpha^{(n)*}(\mathbf{r}_2)\right\rangle - \left|\langle \mathbf{j}^{(n)}\rangle\right|^2) \mathrm{d}\mathbf{r}_1 d\mathbf{r}_2$, where $k = \omega/c$ and the angular brackets stand for the ensemble average. As in [155], we assume that the integrand vanishes for distances $r \ll \lambda$, where $\mathbf{r} = \mathbf{r}_2 - \mathbf{r}_1$; therefore, we omit the $\sim \exp(i\mathbf{k}\cdot\mathbf{r})$ term. This term, however, must be kept for a homogeneous film; in this case, the integration of $\exp(i\mathbf{k}\cdot\mathbf{r}) = \exp(ikr\cos\Theta)$ in the integrand of S over r results in $\delta(\cos\Theta)$, which is non-zero only for the reflected and transmitted direction (we consider here the case of normal incidence).

Using the above expression for $\mathbf{j}^{(n)}(\mathbf{r})$, we can write the current correlator as

$$\int \left\langle j_\alpha^{(n)}(\mathbf{r}_1)\, j_\alpha^{(n)*}(\mathbf{r}_2)\right\rangle d\mathbf{r}_1 d\mathbf{r}_2$$

$$= \int \left\langle \Sigma_{\gamma\beta}^{(n)}(\mathbf{r}_1,\mathbf{r}_3)\Sigma_{\delta\alpha}^{(n)*}(\mathbf{r}_2,\mathbf{r}_4)\delta_{\gamma\delta}\left\{E_\beta^{(n)}(\mathbf{r}_3)E_\alpha^{*(n)}(\mathbf{r}_4)\right\}\right\rangle \prod_{i=1}^{4} d\mathbf{r}_i, \quad (5.73)$$

where $\{\cdots\}$ denotes the averaging over light polarizations. For the unpolarized light, we have $\delta_{\gamma\delta} = 2\left\{E_{n\omega,\,\gamma}^{(0)} E_{n\omega,\,\delta}^{(0)*}\right\}/|E_{n\omega}^{(0)}|^2$, where $\mathbf{E}_{n\omega}^{(0)}$ is the amplitude of a uniform "probe" field at a frequency $n\omega$. We substitute this result and formula (5.73) in the expression for S and integrate over coordinates $\mathbf{r}_1$ and $\mathbf{r}_2$. Then, using the symmetry of the nonlocal conductivity $\Sigma_{\alpha\beta}(\mathbf{r}',\mathbf{r}'') = \Sigma_{\beta\alpha}(\mathbf{r}'',\mathbf{r}')$, we obtain

$$S = \frac{8\pi k^2}{3c|E_{n\omega}^{(0)}|^2}\left|\frac{\sigma^{(n)}}{\sigma^{(1)}}\right|^2 A\left\langle |\sigma E_{n\omega}|^2\,|E_\omega|^{2n}\right\rangle \int_0^\infty g(r)r\,dr, \quad (5.74)$$

where $\mathbf{E}_{n\omega}$ is the local $n\omega$ field excited in the film by the "probe" field $\mathbf{E}_{n\omega}^{(0)}$, σ is the film conductivity at frequency $n\omega$, A is the area of the film, and $g(r)$ is the correlation function defined as

$$g(r) = \left[\left\langle \sigma(\mathbf{r}_1)\sigma^*(\mathbf{r}_2)\left(\mathbf{E}_{n\omega}(\mathbf{r}_1)\cdot\mathbf{E}_{n\omega}^*(\mathbf{r}_2)\right)\left(\mathbf{E}^{(n)}(\mathbf{r}_1)\cdot\mathbf{E}^{(n)*}(\mathbf{r}_2)\right)\right\rangle\right.$$
$$\left. - \left|\left\langle \sigma\left(\mathbf{E}^{(n)}\cdot\mathbf{E}_{n\omega}\right)\right\rangle\right|^2\right]\frac{1}{\left\langle |\sigma\mathbf{E}_{n\omega}|^2\,|\mathbf{E}^{(n)}|^2\right\rangle}. \quad (5.75)$$

(Note that for simplicity we use hereafter $|E_\omega|^{2n}$ instead of the correct expression $|\mathbf{E}_\omega|^2|E_\omega|^{2(n-1)}$.)

We compare this PENS with the $n\omega$ signal $I_{n\omega}$ from the nonlinear layer on a dielectric film with no metal grains on it, where $I_{n\omega}$therefore equals $(c\epsilon_{\rm d}^2/2\pi)\,A\left|\sigma^{(n)}/\sigma^{(1)}\right|^2\left|E_\omega^{(0)}\right|^{2n}$. By expressing the enhancement factor for PENS, $G^{(n)} = S/I_{n\omega}$, in terms of the local dielectric constant ϵ at the frequency $n\omega$, we obtain [175]

$$G^{(n)} = \frac{1}{3}(ka)^4\,\frac{\left\langle |\epsilon E_{n\omega}|^2\,|E_\omega|^{2n}\right\rangle}{\epsilon_{\rm d}^2\left|E_{n\omega}^{(0)}\right|^2\left|E_\omega^{(0)}\right|^{2n}}\,\frac{n^2}{a^2}\int_0^\infty g(r)r\,dr. \quad (5.76)$$

Note that, for a homogeneous ($p = 0$ and $p = 1$) surface, the scattering occurs in the reflected direction only; as mentioned above, the integration of the term $\exp(i\mathbf{k} \cdot \mathbf{r})$, in this case, results in the nonlinear wave propagating in the reflected direction. Besides the small factor $(ka)^4$, which is similar to that in the standard linear Rayleigh scattering, the enhancement $G^{(n)}$ for PENS is proportional to the $2(n+1)$ power of the local field E: $G^{(n)} \sim \left\langle |E|^{2(n+1)} \right\rangle$; more exactly, it is proportional to $G^{(n)} \sim \langle |\epsilon_{n\omega} E_{n\omega}|^2 |E_\omega|^{2n} \rangle$. For highly fluctuating local fields, this factor can be very large (see Fig. 5.12). This enhancement is much larger than that for coherent nonlinear scattering, resulting in a wave that propagates in the reflected direction with the enhancement proportional to $|\langle \epsilon_{n\omega} E_{n\omega} E_\omega^n \rangle|^2$.

The field correlation function $g(r)$ is shown in Fig. 5.13. The function drops very rapidly for $r > a$, and has a negative minimum regardless of the magnitude of field correlation length ξ_e; the anticorrelation occurs because the field maxima have different signs, as seen in Fig. 5.12. The power-law decrease of $g(r)$, which is typical for critical phenomena, occurs in the tail only (see inset in Fig. 5.13). We can speculate that $g(r)$ deviates from the power-law (the straight line in Fig. 5.13) for $r > \xi_e$; using (5.50) for ξ_e, we find $\xi_e(\lambda) \simeq 5$, 20 and 30 (in a units) for $\lambda = 0.34\,\mu\text{m}$, $0.53\,\mu\text{m}$ and $0.9\,\mu\text{m}$ respectively (see Fig. 5.13). For a typical size of a metal grain in a semicontinuous film $a \sim 10$ nm, the intrinsic spatial scale of the local field

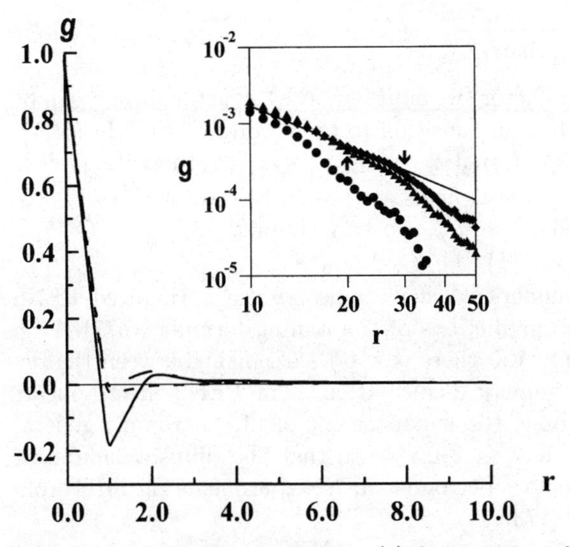

Fig. 5.13. Correlation function $g(r)$ for $n = 3$ at $\lambda_1 = 0.34\,\mu\text{m}$ (*solid line and circles, in the inset*), $\lambda_2 = 0.53\,\mu\text{m}$ (*dashed line and triangles*), and $\lambda_3 = 0.9\,\mu\text{m}$ (*point-dashed line and diamonds*). The arrows are theoretical estimates for $\xi_e(\lambda_2)$ and $\xi_e(\lambda_3)$, and the straight line illustrates the scaling dependence of $g(r)$ in the tail. The units used are those in which $a = 1$

inhomogeneity ξ_e is less than λ, as required by the quasistatic approximation used in the calculations. Note that for a Drude metal and $\omega \ll \omega_p$, the quasistatic approximation holds when the parameter $(a/\lambda)\omega_p/\sqrt{\omega\omega_\tau} \ll 1$, which is typically fulfilled for metal semicontinuous films. The integral of $g(r)$ in (5.76) is estimated as unity for all frequencies.

From the spatial behavior of $g(r)$ and the field distribution shown in Fig. 5.12, we anticipate that, in contrast to harmonic generation from "conventional" metal surfaces [172, 173], the PENS is characterized by a broad-angled distribution with the integral (over all directions) scattering much larger than the coherent scattering in the reflected direction.

To estimate PENS quantitatively, we take into account that $l_r(\omega)$ at the fundamental frequency ω is significantly larger than $l_r(n\omega)$ at the generated frequency $n\omega$. Therefore we can decouple the average $\left\langle |\epsilon \mathbf{E}_{n\omega}|^2 |E_\omega|^{2n} \right\rangle$in (5.76) and approximate it by $\langle |\epsilon E_{n\omega}|^2 \rangle \langle |E_\omega|^{2n} \rangle$, where the second moment of the current is estimated as $\langle |\epsilon E_{n\omega}|^2 \rangle \sim |E_{n\omega}^{(0)}|^2 |\bar{\epsilon}_m(l_r)|^2_{n\omega} M_{2,0}(n\omega) = |E_{n\omega}^{(0)}|^2 |\epsilon_m(n\omega)| \epsilon_d M_{2,0}(n\omega)$. (Note that the latter equality is exact for a self-dual system and can be obtained using the arguments of the paper by Dykhne [147].) Finally, using (5.76) and (5.54), we estimate the PENS factor $G^{(n)}$for the nth harmonic as follows [175]

$$\frac{G^{(n)}}{(ka)^4} \simeq C \frac{n^2}{\epsilon_d} |\epsilon_d \epsilon_m(n\omega)| M_{2,0}(n\omega) M_{2n,0}(\omega)$$

$$\simeq C n^2 \frac{|\epsilon_m(n\omega)|^{5/2} |\epsilon_m(\omega)|^{3(n-1/2)}}{\epsilon_d^{n+1} \epsilon_m''(n\omega) \epsilon_m''(\omega)^{2n-1}}, \tag{5.77}$$

where C is an adjustable pre-factor (we omit here the ξ_A factor since it can be included within C). Note that in transition to the second relation in (5.77), it was assumed that $n\omega < \omega_p$, so that $\epsilon_m'(n\omega)$ is negative; otherwise, $G^{(n)} \simeq C(ka)^4 M_{2n,0}(\omega)$ since the local $n\omega$ fields are not enhanced for $\epsilon_m'(n\omega) > 0$.

For a Drude metal and $\omega \ll \omega_p$, we can simplify (5.77) as $G^{(n)} \simeq C(ka)^4 (\omega_p/\omega_\tau)^{2n} (\omega_p/\omega)^2 / (\epsilon_d^{n+1})$, i.e. $G^{(n)} \propto \lambda^{-2}$.

Figure 5.14 compares numerical calculations for the normalized PENS factors $G^{(n)}/(ka)^4$ with the predictions of the scaling formula (5.77) [175]. For a very large spectral interval, there is good agreement between the developed scaling theory and numerical calculations. The PENS effect appears to be really huge; for example, the enhancement of fifth-harmonic generation is $G^{(5)}/(ka)^4 \sim 10^{21}$ at $\lambda \sim 1\,\mu$m. Note that the diffusive nonlinear scattering was observed by Aktsipetrov et al. in experiments on SHG from semicontinuous metal films [173].

Thus, large field fluctuations in random metal-dielectric composites result in strongly enhanced optical nonlinearities and nonlinear light scattering.

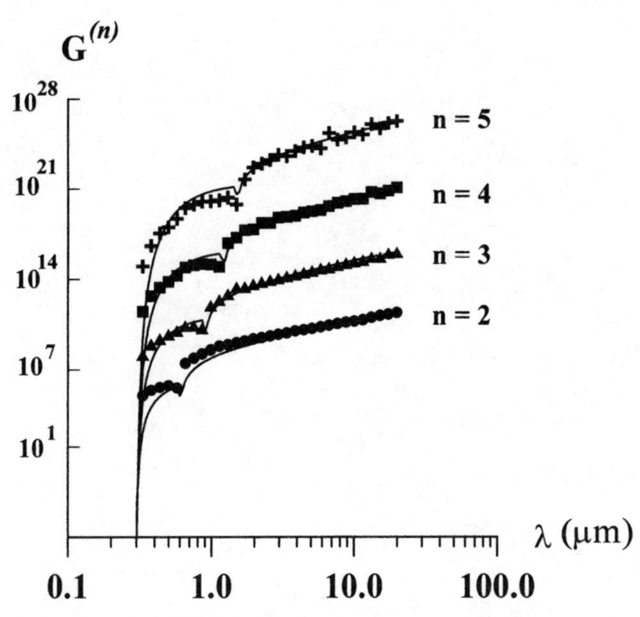

Fig. 5.14. PENS factor $G^{(n)}$ (normalized by $(ka)^4$) for the nth harmonic generation in a silver semicontinuous film at $p = p_c$. Numerical calculations for $n = 2, 3, 4$ and 5 are represented by bullets, triangles, squares and crosses respectively. The solid lines describe $G^{(n)}$ found from the scaling formula (5.77)

References

1. *Electron Transport and Optical Properties of Inhomogeneous Media*, Eds: J. C. Garland and D. B. Tanner, AIP, New York (1978); *Electron Transport and Optical Properties of Inhomogeneous Media (ETOPIM 3)*, Eds: W. L. Mochan and R. G. Barrera, North-Holland, Amsterdam (1994); Physica A **241**, Nos. 1-2, pp. 1-452 (1997), Proceedings of the Fourth International Conference on Electrical Transport and Optical Properties of Inhomogeneous Media (ETOPIM 4), Eds: A. M. Dykhne, A. N. Lagarkov, A.K. Sarychev
2. B. B. Mandelbrot, *The Fractal Geometry of Nature* Freeman, San Francisco, 1982
3. B. Sapoval, *Fractals* (Aditech, Paris, 1990)
4. A. Bunde and S. Havlin. In: *Fractals and Disordered Systems* (eds. A. Bunde and S. Havlin), Springer Verlag, Heidelberg 1991; *Fractals in Physics*, edited by L. Pietroniero and E. Tosatti (North-Holland, Amsterdam, 1986)
5. S. Alexander and R. Orbach, J. Physique - Lettres **43**, 625 (1982)
6. R. Rammal and G. Toulouse, J. Physique - Lettres **44**, 13 (1983)
7. R. Jullien and R. Botet, *Aggregation and Fractal Aggregates*, World Scientific, Singapore (1987); J. Feder, *Fractals* Plenum Press, New York (1988)
8. D. Weitz and M. Oliveria, Phys. Rev. Lett. **52**, 1433 (1984); J. A. Creighton, Metal colloids, in: *Surface Enhanced Raman Scattering*, edited by R. K. Chang and T. E. Furtak, Plenum Press, New York (1982)
9. N. Lu and C. M. Sorensen, Phys. Rev. E **50**, 3109 (1994); J. Cai, N. Lu, and C. M. Sorensen, J. Colloid Interface Sci. **171**, 470 (1995); E. F. Mikhailov and S. S. Vlasenko, Physics-Uspekhi **165**, 253 (1995)
10. R. Chiarello, V. Panella, J. Krim and C. Thompson, Phys. Rev. Lett. **67**, 3408 (1991); C. Douektis, Z. Wang, T. L. Haslett and M. Moskovits, Phys. Rev. B **51**, 11022 (1995); A. L. Barbasi and H. E. Stanley, *Fractal Concepts in Surface Growth*, Cambridge U. Press, Cambridge (1995)
11. V. M. Shalaev, Phys. Rep. **272**, 61 (1996)
12. T. A. Witten and L. M. Sander, Phys. Rev. Lett. **47**, 1400 (1981)
13. D. Stauffer and A. Aharony, *Introduction to Percolation Theory*, 2 ed., Taylor and Francis, Philadelphia (1991)
14. H. E. Stanley, J. Phys. A **10**, L211 (1977)
15. J. C. M. Garnett, Phil. Trans. R. Soc. L. **203**, 385 (1904); ibid. **205**, 237 (1906)
16. R. Clausius, Mechanishe Warmetheorie, Brounschweig, **2**, 62 (1878)
17. O. F. Mossotti, Mem. Soc. Sci. Modena, **14**, 49 (1850)
18. H. A. Lorentz, Wiedem. Ann. **9**, 641 (1880)
19. L. Lorentz, Wieden. Ann. 11 (1881)
20. D. J. Bergman and D. Stroud. In: Solid State Physics **46**, 147, Academic Press, Inc. (1992)
21. D. A. G. Bruggeman, Ann. Physik. (Leipzig) **24**, 636 (1935)

22. D. J. Bergman, Physics Reports **43**, 377 (1978)
23. D. Stroud, Phys. Rev. B **19**, 1783 (1979)
24. K. D. Cummings, J. C. Garland, and D. B. Tanner, Phys. Rev. B **30**, 4170 (1984); D. B. Tanner, Phys. Rev. B **30**, 1042 (1984); P. N. Sen and D. B. Tanner, Phys. Rev. B **26**, 3582 (1982)
25. S. Kirkpatrick, Reviews of Modern Physics **45**, 574 (1973)
26. J. P. Clerc, G. Girard, J. M. Laugier and J. M. Luck, Advances in Physics **39**, 191 (1990)
27. B. Derrida and J. Vannimenus, J. Phys. A **15**, L557 (1982); B. Derrida, D. Stauffer, H. J. Herrmann, and J. Vannimenus, J. Phys. Lett. **44**, L701 (1983); H. J. Herrmann, B. Derrida, and J. Vannimenus, Phys. Rev. B **30**, 4080 (1984)
28. D. J. Frank and C. J. Lobb, Phys. Rev. B **37**, 302 (1988)
29. D. J. Bergman, Phys. Rev. B **14**, 4304 (1976); D. J. Media, Eds: J. C. Garland and D. B. Tanner, AIP, New York, 46 (1978)
30. R. Fuchs, Phys. Rev. B **11**, 1732 (1975)
31. K. Ghosh and R. Fuchs, Phys. Rev. B **38**, 5222 (1988)
32. R. Fuchs, F. Claro, Phys. Rev. B **39**, 3875 (1989)
33. K. Ghosh, R. Fuchs, Phys. Rev. B **44**, 7330 (1991)
34. F. Claro, R. Fuchs, Phys. Rev. B **44**, 4109 (1991)
35. G. Milton, J. Appl. Phys. **52**, 5286 (1981); D. Stroud, G. W. Milton, and B. R. De, Phys. Rev. B **34**, 5145 (1986)
36. H. Ma, W. Wen, W.Y. Tam, and Ping Sheng, Phys. Rev. Lett. **77**, 2499 (1996)
37. A. L. Efros and B. I. Shklovskii, Phys. Stat. Sol. **76**, 475 (1976)
38. J. P. Straley, J. Phys. C: Solid State Phys. **9**, 783 (1976)
39. D. Stroud and D. Bergman, Phys. Rev. B **25**, 2061 (1982)
40. Y. Yagil, M. Yosefin, D. J. Bergman, G. Deutscher, P. Gadenne, Phys. Rev. B **43**, 11342 (1991)
41. P. R. Devaty, Phys. Rev. B **44**, 593 (1991)
42. X. Zhang and D. Stroud, Phys. Rev. B **48**, 6658 (1993)
43. G. A. Niklasson, J. Appl. Phys. **62**, R1 (1987); Physica D **38**, 260 (1989)
44. Y. Gefen, A. Aharony, and S. Alexander, Phys. Rev. Lett. **50**, 77 (1983)
45. V. M. Shalaev and M. I. Stockman, Sov. Phys. JETP **65**, 287 (1987); A. V. Butenko, V. M. Shalaev, and M. I. Stockman, Sov. Phys. JETP **67**, 60 (1988); Z. Phys. D-Atoms, Molecules and Clusters, **10**, 71 (1988); Z. Phys. D - Atoms, Molecules and Clusters, **10**, 81 (1988)
46. V. A. Markel, L. S. Muratov, M. I. Stockman and T. F. George, Phys. Rev. **B43**, 8183 (1991)
47. D. P. Tsai, J. Kovacs, Z. Wang, M. Moskovits, V. M. Shalaev, J. Suh and R. Botet, Phys. Rev. Lett. **72**, 4149 (1994); V. M. Shalaev and M. Moskovits, Phys. Rev. Lett. **75**, 2451 (1995); P. Zhang, T. L. Haslett, C. Douketis, and M. Moskovits, Phys. Rev. B **57**, 15513 (1998)
48. M. I. Stockman, L. N. Pandey, L. S. Muratov and T. F. George, Phys. Rev. Lett. **72**, 2486 (1994); M. I. Stockman, L. N. Pandey, L. S. Muratov and T. F. George, Phys. Rev. B **51**, 185 (1995); M. I. Stockman, L. N. Pandey and T. F. George, Phys. Rev. B **53**, 2183 (1996)
49. M.I. Stockman, Phys. Rev. E **56**, 6494 (1997); Phys. Rev. Lett. **79**, 4562 (1997)
50. V. P. Safonov, V. M. Shalaev, V. A. Markel, Yu. E. Danilova, N. N. Lepeshkin, W. Kim, S. G. Rautian, and R. L. Armstrong, Phys. Rev. Lett. **80**, 1102 (1998)
51. M. Moskovits, Rev. Mod. Phys. **57** , 783 (1985); *Surface Enhance Raman Scattering*, eds. R. K. Chang and T. E. Furtak (Plenum Press, New York, 1982); A. Otto, I. Mrozek, H. Grabhorn, W. Akemann, J. Phys.: Condensed Matter **4**, 1143 (1992)

52. M. I. Stockman, V. M. Shalaev, M. Moskovits, R. Botet and T.F. George, Phys. Rev. B **46**, 2821 (1992)
53. V. A. Markel, V. M. Shalaev, E. B. Stechel, W. Kim and R. L. Armstrong, Phys. Rev. B **53**, 2425 (1996)
54. V. M. Shalaev, E. Y. Poliakov and V. A. Markel, Phys. Rev. B **53**, 2437 (1996)
55. R. W. Boyd, *Nonlinear Optics*, Academic Press, New York (1992); L. D. Landau, E. M. Lifshits and L. P. Pitaevskii, *Electrodynamics of Continuous Media*, 2nd edn, Pergamon, Oxford (1984)
56. H.D. Bale and P. W. Schmidt, Phys. Rev. Lett. **53**, 596 (1984)
57. M. V. Berry and I. C. Percival, Optica Acta **33**, 577 (1986)
58. J. E. Martin and A. J. Hurd, J. Appl. Cryst. **20**, 61 (1987)
59. N. Lu and C. M. Sorensen, Phys. Rev. E **50**, 3109 (1994); J. Cai, N. Lu, and C. M. Sorensen, J. Colloid Interface Sci. **171**, 470 (1995)
60. V. A. Markel, V. M. Shalaev, E. Yu. Poliakov, and T. F. George, Phys. Rev. E **55**, 7313 (1997); J. Opt. Soc. Am A **14**, 60 (1997); V. A. Markel, V. M. Shalaev, E. Y. Poliakov, T. F. George, in: *Fractal Frontiers*, Eds: M. M. Novak and T. G. Dewey, World Scientific, Singapore, 1997; p. 291
61. M. Carpineti, M. Giglio, and V. Deriorgio, Phys. Rev. E **51**, 590 (1995)
62. F. Sciortino, A. Belloni, and P. Tartaglia, Phys. Rev. E **52**, 4068 (1995)
63. N. G. Khlebtsov and A. G. Mel'nikov, Opt. Spectrosc. (USSR) **79**, 656 (1995); N. G. Khlebtsov, Colloid J. **58**, 100 (1996)
64. S. D. Andreev, L. S. Ivlev, E. F. Mikhailov, and A. A. Kiselev, Atmos. Oceanic Opt. **8**, 355 (1995)
65. S. Alexander, Phys. Rev. B **40**, 7953 (1989)
66. V. M. Shalaev, R. Botet, and R. Julien, Phys. Rev. B **44**, 12216 (1991)
67. M. I. Stockman, T. F. George, and V. M. Shalaev, Phys. Rev. B **44**, 115 (1991)
68. V. M. Shalaev, M. I. Stockman, and R. Botet, Physica A **185**, 181 (1992)
69. V. M. Shalaev, V. A. Markel, V. P. Safonov, Fractals **2**, 201 (1994); V. M. Shalaev, R. Botet, D. P. Tsai, J. Kovacs, M. Moskovits, Physica A **207**, 197 (1994)
70. V. M. Shalaev, V. A. Markel, E. Y. Poliakov, R. L. Armstrong, V.P. Safonov, A. K. Sarychev, J. Nonlinear Optic. Phys. and Materials, v. 7, 131 (1998); V. M. Shalaev, E. Y. Poliakov, V. A. Markel, V. P. Safonov, A. K. Sarychev, Fractals, **5** (suppl.), 63 (1997); Vladimir M. Shalaev, E.Y. Poliakov, V.A. Markel, and R. Botet, Physica A **241**, 249 (1997)
71. Vladimir M. Shalaev, V. P. Safonov, E.Y. Poliakov, V. A. Markel, and A. K. Sarychev, Fractal-Surface-Enhanced Optical Nonlinearities, in: *Nanostructured Materials: Clusters, Composites, and Thin Films*, Eds; Vladimir M. Shalaev and Martin Moskovits, ACS Symposium Series v. 679, ACS Books, 1997; V. M. Shalaev, E. Y. Poliakov, V. A. Markel, R. Botet, Physica A **241**, 249 (1997); R. Botet, E. Y. Poliakov, V. M. Shalaev, and V. A. Markel, Fractal-Surface-Enhanced Optical Responses, in: *Fractals in Engineering*, Eds: J. Levy Vehel, E. Lutton and Claude Tricot, Springer-Verlag, London (1997); p. 237
72. Vladimir M. Shalaev, Surface-Enhanced Optical Phenomena in Nanostructured Fractal Materials, in *Handbook on Nanostructured Materials and Nanotechnology*, ed. H. S. Nalwa, (Academic Press, 1999)
73. S.G. Rautian, V.P. Safonov, P.A. Chubakov, V.M. Shalaev, M.I. Shtockman, JETP Lett. **47**, 243 (1988) [transl. from Pis'ma Zh.Eksp.Teor.Fiz. **47**, 200 (1988)]
74. Yu. E. Danilova, V. P. Drachev, S. V. Perminov and V. P. Safonov, Bulletin of the Russian Acad. Sci., Physics, **60**, 342 (1996); Ibid., 374; Yu. E. Danilova, N. N. Lepeshkin, S. G. Rautian and V. P. Safonov, Physica A **241**, 231 (1997); F.

A. Zhuravlev, N. A. Orlova, V. V. Shelkovnikov, A. I. Plehanov, S. G. Rautian and V. P. Safonov, *JETP Lett.* **56**, 260 (1992)
75. V. M. Shalaev, R. Botet, A. V. Butenko, Phys. Rev. B **48**, 6662 (1993)
76. V. M. Shalaev, R. Botet, Phys. Rev. B **50**, 12987 (1994)
77. S. I. Bozhevolnyi, V. A. Markel, V. Coello, W. Kim, and V. M. Shalaev, Phys. Rev. B **58**, 11 441 (1998)
78. A. V. Butenko, P. A. Chubakov, Yu. E. Danilova, S. V. Karpov, A. K. Popov, S. G. Rautian, V. P. Safonov, V. V. Slabko, V. M. Shalaev, and M. I.Stockman, Z. Phys. D- Atoms, Molecules, and Clusters **17**, 283 (1990)
79. S. V. Karpov, A. K. Popov, S. G. Rautian, V. P. Safonov, V. V. Slabko, V. M. Shalaev, M. I. Shtockman, JETP Lett. **48**, 571 (1988) [transl. from Pis'ma Zh.Eksp.Teor.Fiz. **48**, 528 (1988)]
80. V. A. Markel, J. Opt. Soc. Am. B **12**, 1783 (1995); V. A. Markel, J. Mod. Opt. **39**, 853 (1992)
81. E. M. Purcell, C. R. Pennypacker, Astrophys. J. **186**, 705 (1973)
82. B. T. Draine, Astrophys. J. **333**, 848 (1988)
83. A. Lakhtakia, Int. J. Mod. Phys. **3**, 583 (1992); Int. J. Infrared Millimeter Waves **13**, 869 (1992)
84. B. T. Draine and J. Goodman, Astrophys. J. **405**, 685 (1993)
85. J. M. Gerardy and M. Ausloos, Phys. Rev. B **22**, 4950 (1980)
86. F. Claro, Phys. Rev. B **25**, 7875 (1982)
87. J. E. Sansonetti and J. K. Furdyna, Phys. Rev. B **22**, 2866 (1980)
88. V. A. Markel and V. M. Shalaev, Computational Approaches in Optics of Fractal Clusters, in: Computational Studies of New Materials, eds: D. A. Jelski and T. F. George (World Scientific, Singapore, 1999)
89. P. B. Johnson and R. W. Christy, Phys. Rev. B **6**, 4370 (1972)
90. N. M. Lawandy et al., Nature **368**, 436 (1994); S. John and G. Pang, Phys. Rev. A **54**, 3642 (1996); R. M. Balachandran, N. M. Lawandy, J. A. Moon, Opt. Lett. **22**, 319 (1997); V. S. Letokhov, Soviet Physics JETP **26**, 835 (1968); M. A. Noginov et al., J. Opt. Soc. Am. B **14**, 2153 (1997); D. S. Wiersma and A. Lagendijk, Phys. Rev. E **54**, 4256 (1996); G. A. Berger, M. Kempe, and A. Z. Genack, Phys. Rev. E **56**, 6118 (1997)
91. E. M. Purcell, Phys. Rev. **69**, 681 (1946); S. Haroche and D. Klepper, Phys. Today **42**, # 1, 24 (1989); *Microcavities and Photonic Bandgaps: Physics and Applications*, edited by C. Weisbuch and J. Rarity, NATO ASI, ser. E, vol. 324 (Kluwer, Dordrecht, 1996); E. Yablonovitch, *Light Emission in Photonic Crystal Micro-Cavities*, p. 635; in: *Confined Electrons and Photons*, edited by E. Burstein and C. Weisbuch (Plenum Press, New York, 1995)
92. K. Kneipp et al., Phys. Rev. Lett. **78**, 1667 (1997)
93. S. Nie and S. R. Emory, Science **275**, 1102 (1997)
94. P. Gadenne, F. Brouers, V. M. Shalaev, A. K. Sarychev, J. Opt. Soc. Am. B **15**, 68 (1998)
95. V. A. Markel, V. M. Shalaev, P. Zhang, W. Huynh, L. Tay, T. L. Haslett, and M. Moskovits, Phys. Rev. B **59**, 10903 (1999)
96. C. Flytzanis, Prog. Opt. **29**, 2539 (1992); D. Ricard, Ph. Roussignol and C. Flytzanis, Opt. Lett. **10**, 511 (1985); F. Hache, D. Ricard, C. Flytzanis and U. Kreibig, Applied Physics A **47**, 347 (1988)
97. D. Stroud, P. M. Hui, Phys. Rev. B **37**, 8719 (1988); P. M. Hui, D. Stroud, Phys. Rev B **49**, 11729 (1994); D. Stroud, X. Zhang Physica A **207**, 55 (1994); X. Zhang and D, Stroud, Phys. Rev. B **49**, 944 (1994); P. M. Hui, Phys. Rev. B **49**, 15344 (1994); K. W. Yu, Y. C. Chu, and Eliza M. Y. Chan, Phys. Rev. B **50**, 7984 (1994); P. M. Hui, P. Cheung, and Y. R. Kwong, Physica A **241**, 301 (1997)

98. O. Levy, D. J. Bergman, D. G. Stroud, Phys. Rev. E **52**, 3184 (1995); D. Bergman, O. Levy, D. Stroud, Phys. Rev. B **49**, 129 (1994); O. Levy, D. Bergman, Physica A **207**, 157 (1994)
99. W. M. V. Wan, H. C. Lee, P. M. Hui, and K. W. Yu, Phys. Rev. B **54**, 3946 (1996); K. W. Yu, Y. C. Wang, P. M. Hui, G. Q. Gu, Phys. Rev. B **47**, 1782 (1993); K. W. Yu. P. M. Hui, D. Stroud, Phys. Rev. B **47**, 14150 (1993); K. W. Yu, Phys. Rev. B **49**, 9989 (1994); H. C. Lee, K. P. Yuen, and K. W. Yu, Phys. Rev. B **51**, 9317 (1995); L. Fu, L. Resca, Phys. Rev. B **56**, 10963 (1997)
100. H. Ma, R. Xiao, P. Sheng, J. Opt. Soc. Am B **15**, 1022 (1998); H. B. Liao, R. F. Fiao, J. S. Fu, P. Yu, G. K. L. Wong, P. Sheng, Appl. Phys. Lett. **70**, 1 (1997); K. P. Yuen, M. F. Law, K. W. Yu, P. Sheng, Phys. Rev. E **56**, R1322 (1997); H. B. Liao, R. F. Xiao, J. S. Fu, H. Wang, K. S. Wong, G. K. L. Wong, Opt. Lett. **23**, 388 (1988); H. B. Liao, R. F. Xiao, H. Wang, K. S. Wong, and G. K. L. Wong, Appl. Phys. Lett. **72**, 1817 (1998)
101. J. E. Sipe and R. W. Boyd, Phys. Rev. B **46**, 1614 (1992); R. J. Gehr, G. L. Fisher, R. W. Boyd and J. E. Sipe, Phys. Rev. A **53**, 2792 (1996); G. L. Fisher, R. W. Boyd, R. J. Gehr, S. A. Jenekhe, J. A. Osaheni, J. E. Sipe and L. A. Weller-Brophy, Phys. Rev. Lett. **74**, 1871 (1995); R. W. Boyd, R. J. Gehr, G. L. Fisher and J. E. Sipe, Pure Appl. Opt. **5**, 505 (1996)
102. J. S. Suh and M. Moskovits, J. Phys. Chem. **58**, 5526 (1984)
103. P. W. Anderson, Phys. Rev. **109**, 1492 (1958); S. John, Physics Today, 32 (May, 1991); Ad Lagendijk, Bar A. van Tiggelen, Physics Reports **270**, 143 (1996); A. Z. Genack, Phys. Rev. Lett. **58**, 2043 (1987); E. Yablonovich, Phys. Rev. Lett. **58**, 2043 (1987); C. M. Soukoulis, S. Datta, E. N. Economou, Phys. Rev. B **49**, 3800 (1994); P. W. Brouwer, P. G. Silvestrov, and C. W. J. Beenakker, Phys. Rev. B **56**, R4333 (1997)
104. V. M. Shalaev, M. Moskovits, A. A. Golubentsev, and S. John, Physica A **191**, 352 (1992)
105. M. Sheik-Bahae, A. A. Said, T. H. Wei, D. J. Hagan and E. W. VanStryland, IEEE J. Quant. Electr. **Q26**, 760 (1990)
106. S. M. Heard, F. Griezer, C. G. Barrachough, and J. V. Sandera, J. Colloid Interface Sci. **93**, 545 (1983)
107. P. C. Lee and D. Meisel, J. Phys. Chem. **86**, 3391 (1982)
108. W. D. Bragg, V. P. Safonov, W. Kim, K. Banerjii, M. R. Young, J. G. Zhu, Z. C. Ying, R. L. Armstrong, and V. M. Shalaev, J. Microscopy **194**, 574 (1999)
109. R. K. Chang and A. J. Campillo, Ed., *Optical Processes in Microcavities*, (World Scientific, Singapore-New Jersey-London- Hong Kong, 1996)
110. W. Kim, V. P. Safonov, V. M. Shalaev, R. L. Armstrong, Phys. Rev. Lett. **82**, 4811 (1999)
111. H.-M. Tzeng, K. F. Wall, M. B. Long, R. K. Chang, Opt. Lett. **9**, 499 (1984)
112. U. Kreibig and M. Vollmer, *Optical Properties of Metal Clusters* (Springer-Verlag, Berlin Heidelberg, 1995)
113. S. V. Karpov, et al., JETP Lett. **66**, 106 (1997)
114. V. M. Shalaev, R. Botet, J. Mercer, and E. B. Stechel, PRB **54**, 8235 (1996)
115. E. Y. Poliakov, V. M. Shalaev, V. A. Markel, and R. Botet, Opt. Lett. **21**, 1628 (1996); B. Vlckova, C. Douketis, M. Moskovits, V. M. Shalaev, V. A. Markel, J. Chem. Phys. **110**, 8080 (1999)
116. E. Y. Poliakov, V. A. Markel, V. M. Shalaev, R. Botet, Phys. Rev. B **57**, 14901 (1998); V. M. Shalaev, C. Douketis, T. Haslett, T. Stuckless, and M. Mosovits, Phys. Rev. B **53**, 11193 (1996)
117. J. M. Kim and J. M. Kosterlitz, Phys. Rev. Lett. **62**, 2289 (1989)
118. The complete set of formulas for the case of electromagnetic waves scattering from smooth random surfaces can be found in [119, 120]

119. A. Marvin, T. Toigo, and V. Celli, Phys. Rev. B **11**, 2777 (1975); F. Toigo, A. Marvin, V. Celli, and N. Hill, Phys. Rev. B **15**, 5618 (1977)
120. A. A. Maradudin and D. L. Mills Phys. Rev. B **11**, 1392 (1975)
121. P. Beckmann and A. Spizzichino, *The Scattering of Electromagnetic Waves from Rough Surfaces*, Artech, Norwood, Mass. (1987)
122. A. Ishimaru, in: *Scattering in Volumes and Surfaces*, ed. by M. Nieto-Vesperinas and J. C. Dainity, Elsever Sci. Pub., North-Holland (1990)
123. R. M. Fitzgerland and A. A. Maradudin, Waves in Random Media **4**, 275 (1994)
124. E. Betzig, et al., Science **257**, 189 (1992); J.-J. Greffet and R. Carminati, Progress in Surface Science **56**, 133 (1997)
125. A. Petri, L. Pietronero, Phys. Rev. B **45**, 12864 (1992)
126. Y. R. Shen, *The Principles of Nonlinear Optics*, Wiley, New York (1984)
127. G. L. Richmond, J. M. Robinson, and V.L. Shanon, Prog. Surf. Sci. **28**, 1 (1988)
128. S. Janz and H. M. van Driel, Int. J. Nonlinear Opt. Phys. **2**, 1 (1993)
129. J. E. Sipe, V. C. Y. So, M. Fukui, and G. I. Stegeman, Phys. Rev. B **21**, 4389 (1980); P. Guyot-Sionnest, W. Chen, and Y. R. Shen, Phys. Rev. B **33**, 8254 (1986)
130. R. Murphy, M. Yeganeh, K. J. Song, and E. W. Plummer Phys. Rev. Lett. **63**, 318 (1989); O. A. Aktsipetrov, A. A. Nikulin, V. I. Panov, and S. I. Vasil'ev, Solid St. Com. **73**, 411 (1990); H. Ishida and A. Liebsch, Phys. Rev. B **50**, 4834 (1994); A. V. Petukhov and A. Liebsch, Surf. Sci. **334**, 195 (1995)
131. P. S. Pershan Phys. Rev. Lett. **130**, 919 (1963); P. Guyot-Sionnest and Y. R. Shen, Phys. Rev. B **38**, 7985 (1988)
132. B. S. Mendoza, W. L. Mochan, Phys. Rev. B **53**, 4999 (1996)
133. J. Rudnick and E. A. Stern, Phys. Rev. B **4**, 4274 (1971)
134. Note that in this simplified model the orientation of **n** does not change from site to site
135. R.W. Cohen, G.D. Cody, M.D. Coutts and B. Abeles, Phys. Rev. B **8**, 3689 (1973)
136. G. A. Niklasson and C. G. Granqvist, J. Appl. Phys. **55**, 3382 (1984)
137. Y. Yagil, P. Gadenne, C. Julien and G. Deutscher, Phys. Rev. **46**, 2503 (1992)
138. T. W. Noh, P.H. Song, Sung-Il Lee, D. C. Harris, J. R. Gaines and J. C. Garland, Phys. Rev. **46**, 4212 (1992)
139. P. Gadenne, A. Beghadi, and J. Lafait, Optics Comm., **65**, 17 (1988)
140. P. Gadenne, Y. Yagil and G. Deutscher, J. Appl. Phys. **66**, 3019 (1989)
141. Y. Yagil, M. Yosefin, D. J. Bergman, G. Deutscher and P. Gadenne, Phys. Rev. B **43**, 11342 (1991)
142. F. Brouers, J. P. Clerc and G. Giraud, Phys. Rev. B **44**, 5299 (1991); F. Brouers, J. M. Jolet, G. Giraud, J. M.. Laugier, Z. A. Randriamanantany Physica A **207**, 100 (1994)
143. A. P. Vinogradov, A. M. Karimov and A. K. Sarychev Zh. Eksp. Teor. Fiz., **94**, 301 (1988) [Sov. Phys. JETP, **67**, 2129 (1988)]
144. G. Depardieu, P. Frioni and S. Berthier Physica A **207**, 110 (1994)
145. A. Aharony, Phys. Rev. Lett. **58**, 2726 (1987)
146. D. J. Bergman, Phys. Rev. B **39**, 4598 (1989)
147. A. M. Dykhne, Sov. Phys. JETP **32**, 348(1971) [trans. from Zh. Eksp. Teor. Fiz. **59**, 110 (1970)]
148. F. Brouers, S. Blacher, and A. K. Sarychev, in: *Fractals in the Natural and Applied Sciences*, Chapman and Hall, chap. 24, London, 1995; F. Brouers, A. K. Sarychev, S. Blacher, O. Lothaire, Physica A **241**, 146 (1997); F. Brouers, S. Blacher, and A. K. Sarychev, Phys. Rev. B **58**, 15897 (1998)

149. F. Brouers, S. Blacher, A. N. Lagarkov, A. K. Sarychev, P. Gadenne, and V. M. Shalaev, Phys. Rev. B **55**, 13234, (1997); P. Gadenne, F. Brouers, V. M. Shalaev, and A. K. Sarychev, J. Opt. Soc. Am. B **15**, 68 (1998)
150. V. M. Shalaev, A. K. Sarychev, Phys. Rev. B **57**, 13265 (1998)
151. P. J. Reynolds, W. Klein, and H. E. Stanley, J. Phys. C **10**, L167 (1977)
152. A. K. Sarychev, Zh. Eksp. Teor. Fiz. **72**, 1001 (1977) [Sov. Phys. JETP **45**, 524 (1977)]
153. J. Bernasconi, Phys. Rev. B **18**, 2185 (1978)
154. A. Aharony, Physica A **205**, 330 (1994)
155. H. E. Stanley, *Introduction to Phase Transition and Critical Phenomena*, Oxford Press (1981); I. L. Fabelinskii, *Molecular Scattering of Light*, Plenum, NY (1968)
156. P. M. Chaikin and T. C. Lubensky, *Principles of Condensed Matter Physics* (Cambridge Univ. Press, Cambridge, 1995)
157. B. Kramer, A. MacKinnon, Reports on Progress in Physics **56**, 1469 (1993)
158. D. Belitz and T. R. Kirkpatrick, Rev. Mod. Phys. **66**, 261 (1994); M. V. Sadovskii, Physics Reports **282**, 225 (1997)
159. V. I. Fal'ko and K. B. Efetov, Phys. Rev. B **52**, 17413 (1995)
160. K. B. Efetov, *Supersymmetry in Disorder and Chaos* (Cambridge Univ. Press, UK, 1997)
161. M. V. Berry, J. Phys. A **10,** 2083 (1977)
162. A. V. Andreev, et al., Phys. Rev. Lett. **76**, 3947 (1996)
163. K. Muller, et. al., Phys. Rev. Lett. **78**, 215 (1997)
164. J. A. Verges, Phys. Rev. B **57**, 870 (1998)
165. A. Elimes, R. A. Romer and M. Schreiber, Eur. Phys. J. B **1**, 29 (1998)
166. M. Kaveh and N. F. Mott, J. Phys. A **14**, 259 (1981)
167. T. Kawarabayashi, B. Kramer and T. Ohtsuki, Phys. Rev. B **57**, 11842 (1998)
168. S. Gresillon, L. Aigouy, A. C. Boccara, J. C. Rivoal, X. Quelin, C. Desmarest, P. Gadenne, V. A. Shubin, A. K. Sarychev, and V. M. Shalaev, Phys. Rev. Lett. **82**, 4520 (1999)
169. W. D. Bragg, V. Podolskii, K. Banerjee, Z. C. Ying, R. L. Armstrong, A. K. Sarychev, and V. M. Shalaev (unpublished)
170. R. R. Alfano and S. L. Shapiro, Phys. Rev. Lett. **24**, 592 (1970); G. Yang and Y. R. Sheng, Opt. Lett. **9,** 510 (1984); J. T. Manassah, M. A. Mustafa, R. R. Alfano, and P. P. Ho, IEEE J. Quantum Electron. **224**, 197 (1986); P. B. Corkum, C. Rolland, T. Srinivasan-Rao, Phys. Rev. Lett. **57**, 2268 (1986); R. L. Fork, C. V. Shank, C. Hirliman, R. Yen, W. J. Tomlinson, Opt. Lett. **8**, 1 (1983)
171. X. Quelin, C. Desmarest, P. Gadenne, S. Gresillon, A. C. Boccara, J. C. Rivoal, V. A. Shubin, and V. M. Shalaev (unpublished)
172. G. A. Reider, T. F. Heinz, in *Photonic Probes of Surfaces: Electromagnetic Waves,* ed. by P. Halevi, Elsevier, 1995; A. Liebsch, W. L. Schaich, Phys. Rev. B **40**, 5401 (1989); R. Murphy, et al., Phys. Rev. Lett **63**, 318 (1989); T. Y. F. Tsang, Phys. Rev. A **52**, 4116 (1995)
173. T. Y. F. Tsang, Opt. Lett. **21**, 245 (1996); O. A. Aktsipetrov, et al., Solid State Commun. **73**, 411 (1990); A. R. McGurn, Surf. Sci. Rep. **10**, 359 (1990); A. R. McGurn, et. al., Phys. Rev. B **31**, 4866 (1985); O. A. Aktsipetrov, et. al., Phys. Let. A **179**, 149 (1993); L. Kuang and H. J. Simon, Ibid. **197**, 257 (1995)
174. This expression, strictly speaking, holds only for the scalar nonlinear conductivity and odd n (i.e., $n = 2k + 1$), when $E^{n-1} = (\mathbf{E} \cdot \mathbf{E})^k$. However, for estimates, the formula can be used in the general case, for arbitrary n
175. A. K. Sarychev, V. A. Shubin, and V. M. Shalaev, Phys. Rev. E **59**, 7239 (1999)

Index

Springer Tracts in Modern Physics

140 **Exclusive Production of Neutral Vector Mesons at the Electron-Proton Collider HERA**
By J. A. Crittenden 1997. 34 figs. VIII, 108 pages

141 **Disordered Alloys**
Diffusive Scattering and Monte Carlo Simulations
By W. Schweika 1998. 48 figs. X, 126 pages

142 **Phonon Raman Scattering in Semiconductors, Quantum Wells and Superlattices**
Basic Results and Applications
By T. Ruf 1998. 143 figs. VIII, 252 pages

143 **Femtosecond Real-Time Spectroscopy of Small Molecules and Clusters**
By E. Schreiber 1998. 131 figs. XII, 212 pages

144 **New Aspects of Electromagnetic and Acoustic Wave Diffusion**
By POAN Research Group 1998. 31 figs. IX, 117 pages

145 **Handbook of Feynman Path Integrals**
By C. Grosche and F. Steiner 1998. X, 449 pages

146 **Low-Energy Ion Irradiation of Solid Surfaces**
By H. Gnaser 1999. 93 figs. VIII, 293 pages

147 **Dispersion, Complex Analysis and Optical Spectroscopy**
By K.-E. Peiponen, E.M. Vartiainen, and T. Asakura 1999. 46 figs. VIII, 130 pages

148 **X-Ray Scattering from Soft-Matter Thin Films**
Materials Science and Basic Research
By M. Tolan 1999. 98 figs. IX, 197 pages

149 **High-Resolution X-Ray Scattering from Thin Films and Multilayers**
By V. Holý, U. Pietsch, and T. Baumbach 1999. 148 figs. XI, 256 pages

150 **QCD at HERA**
The Hadronic Final State in Deep Inelastic Scattering
By M. Kuhlen 1999. 99 figs. X, 172 pages

151 **Atomic Simulation of Electrooptic and Magnetooptic Oxide Materials**
By H. Donnerberg 1999. 45 figs. VIII, 205 pages

152 **Thermocapillary Convection in Models of Crystal Growth**
By H. Kuhlmann 1999. 101 figs. XVIII, 224 pages

153 **Neutral Kaons**
By R. Belušević 1999. 67 figs. XII, 183 pages

154 **Applied RHEED**
Reflection High-Energy Electron Diffraction During Crystal Growth
By W. Braun 1999. 150 figs. IX, 222 pages

155 **High-Temperature-Superconductor Thin Films at Microwave Frequencies**
By M. Hein 1999. 134 figs. XIV, 395 pages

156 **Growth Processes and Surface Phase Equilibria in Molecular Beam Epitaxy**
By N.N. Ledentsov 1999. 17 figs. VIII, 84 pages

157 **Deposition of Diamond-Like Superhard Materials**
By W. Kulisch 1999. 60 figs. X, 191 pages

158 **Nonlinear Optics of Random Media**
Fractal Composites and Metal-Dielectric Films
By V.M. Shalaev 2000. 51 figs. XII, 158 pages

159 **Magnetic Dichroism in Core-Level Photoemission**
By K. Starke 2000. 64 figs. X, 136 pages

160 **Physics with Tau Leptons**
By A. Stahl 2000. 236 figs. XIII, 313 pages

Printing (computer to plate): Mercedes-Druck, Berlin
Binding: Stürtz AG, Würzburg